T0419833

Gabriel Luque and Enrique Alba

# Parallel Genetic Algorithms

# Studies in Computational Intelligence, Volume 367

**Editor-in-Chief**
Prof. Janusz Kacprzyk
Systems Research Institute
Polish Academy of Sciences
ul. Newelska 6
01-447 Warsaw
Poland
*E-mail:* kacprzyk@ibspan.waw.pl

---

Further volumes of this series can be found on our homepage: springer.com

Vol. 346. Weisi Lin, Dacheng Tao, Janusz Kacprzyk, Zhu Li, Ebroul Izquierdo, and Haohong Wang (Eds.)
*Multimedia Analysis, Processing and Communications,* 2011
ISBN 978-3-642-19550-1

Vol. 347. Sven Helmer, Alexandra Poulovassilis, and Fatos Xhafa
*Reasoning in Event-Based Distributed Systems,* 2011
ISBN 978-3-642-19723-9

Vol. 348. Beniamino Murgante, Giuseppe Borruso, and Alessandra Lapucci (Eds.)
*Geocomputation, Sustainability and Environmental Planning,* 2011
ISBN 978-3-642-19732-1

Vol. 349. Vitor R. Carvalho
*Modeling Intention in Email,* 2011
ISBN 978-3-642-19955-4

Vol. 350. Thanasis Daradoumis, Santi Caballé, Angel A. Juan, and Fatos Xhafa (Eds.)
*Technology-Enhanced Systems and Tools for Collaborative Learning Scaffolding,* 2011
ISBN 978-3-642-19813-7

Vol. 351. Ngoc Thanh Nguyen, Bogdan Trawiński, and Jason J. Jung (Eds.)
*New Challenges for Intelligent Information and Database Systems,* 2011
ISBN 978-3-642-19952-3

Vol. 352. Nik Bessis and Fatos Xhafa (Eds.)
*Next Generation Data Technologies for Collective Computational Intelligence,* 2011
ISBN 978-3-642-20343-5

Vol. 353. Igor Aizenberg
*Complex-Valued Neural Networks with Multi-Valued Neurons,* 2011
ISBN 978-3-642-20352-7

Vol. 354. Ljupco Kocarev and Shiguo Lian (Eds.)
*Chaos-Based Cryptography,* 2011
ISBN 978-3-642-20541-5

Vol. 355. Yan Meng and Yaochu Jin (Eds.)
*Bio-Inspired Self-Organizing Robotic Systems,* 2011
ISBN 978-3-642-20759-4

Vol. 356. Slawomir Koziel and Xin-She Yang (Eds.)
*Computational Optimization, Methods and Algorithms,* 2011
ISBN 978-3-642-20858-4

Vol. 357. Nadia Nedjah, Leandro Santos Coelho, Viviana Cocco Mariani, and Luiza de Macedo Mourelle (Eds.)
*Innovative Computing Methods and their Applications to Engineering Problems,* 2011
ISBN 978-3-642-20957-4

Vol. 358. Norbert Jankowski, Włodzisław Duch, and Krzysztof Grąbczewski (Eds.)
*Meta-Learning in Computational Intelligence,* 2011
ISBN 978-3-642-20979-6

Vol. 359. Xin-She Yang, and Slawomir Koziel (Eds.)
*Computational Optimization and Applications in Engineering and Industry,* 2011
ISBN 978-3-642-20985-7

Vol. 360. Mikhail Moshkov and Beata Zielosko
*Combinatorial Machine Learning,* 2011
ISBN 978-3-642-20994-9

Vol. 361. Vincenzo Pallotta, Alessandro Soro, and Eloisa Vargiu (Eds.)
*Advances in Distributed Agent-Based Retrieval Tools,* 2011
ISBN 978-3-642-21383-0

Vol. 362. Pascal Bouvry, Horacio González-Vélez, and Joanna Kolodziej (Eds.)
*Intelligent Decision Systems in Large-Scale Distributed Environments,* 2011
ISBN 978-3-642-21270-3

Vol. 363. Kishan G. Mehrotra, Chilukuri Mohan, Jae C. Oh, Pramod K. Varshney, and Moonis Ali (Eds.)
*Developing Concepts in Applied Intelligence,* 2011
ISBN 978-3-642-21331-1

Vol. 364. Roger Lee (Ed.)
*Computer and Information Science,* 2011
ISBN 978-3-642-21377-9

Vol. 365. Roger Lee (Ed.)
*Computers, Networks, Systems, and Industrial Engineering 2011,* 2011
ISBN 978-3-642-21374-8

Vol. 366. Mario Köppen, Gerald Schaefer, and Ajith Abraham (Eds.)
*Intelligent Computational Optimization in Engineering,* 2011
ISBN 978-3-642-21704-3

Vol. 367. Gabriel Luque and Enrique Alba
*Parallel Genetic Algorithms,* 2011
ISBN 978-3-642-22083-8

Gabriel Luque and Enrique Alba

# Parallel Genetic Algorithms

## Theory and Real World Applications

**Authors**

Dr. Gabriel Luque
E.T.S.I. Informática
University of Málaga
Campus de Teatinos
29071 Málaga
Spain
Email: gabriel@lcc.uma.es

Prof. Enrique Alba
E.T.S.I. Informática
University of Málaga
Campus de Teatinos
29071 Málaga
Spain
Email: eat@lcc.uma.es

ISBN 978-3-642-22083-8 e-ISBN 978-3-642-22084-5

DOI 10.1007/978-3-642-22084-5

Studies in Computational Intelligence ISSN 1860-949X

Library of Congress Control Number: 2011930858



*Typeset & Cover Design:* Scientific Publishing Services Pvt. Ltd., Chennai, India.

Printed on acid-free paper

9 8 7 6 5 4 3 2 1

springer.com

To my family
*Gabriel Luque*

To my family, for their
continuous support
*Enrique Alba*

# Preface

This book is the result of several years of research trying to better characterize parallel genetic algorithms (pGAs) as a powerful tool for optimization, search, and learning.

We here offer a presentation structured in three parts. The first one is targeted to the algorithms themselves, discussing their components, the physical parallelism, and best practices in using and evaluating them.

A second part deals with theoretical results relevant to the research with pGAs. Here we stress several issues related to actual and common pGAs.

A final third part offers a very wide study of pGAs as problem solvers, addressing domains such as natural language processing, circuits design, scheduling, and genomics. With such a diverse analysis, we intend to show the big success of these techniques in Science and Industry.

We hope this book will be helpful both for researchers and practitioners. The first part shows pGAs to either beginners or researchers looking for a unified view of the field. The second part partially solves (and also opens) new investigation lines in theory of pGAs. The third part can be accessed independently for readers interested in those applications. A small note on MALLBA, one of our software libraries for parallel GAs is also included to ease laboratory practices and actual applications.

We hope the reader will enjoy the contents as much we did in writing this book.

Málaga,
May 2010

Gabriel Luque
Enrique Alba

# Contents

# Part I

# Introduction

# 1

# Introduction

*Be brief in your reasonings, there is no relish in long ones.*

*Miguel de Cervantes (1547 - 1616) - Spanish writer*

Research in exact algorithms, heuristics, and metaheuristics for solving combinatorial optimization problems is nowadays highly on the rise. The main advantage of using exact methods is the guarantee of finding the global optimum for the problem [1], but the critical disadvantage when solving real problems (NP-hard) comes from of the exponential growth of the execution time according to the instance size, as well as from the unreal constraints they often impose to solve the problem. On the other hand, specific (*ad-hoc*) heuristics tend to be very fast [2], but the solutions obtained are generally not of high quality and difficult to export to other similar problems. In contrast, metaheuristics offer a tradeoff between both [3, 4]: they are generic techniques which offer a good solution (even the global optimum) usually in a moderate run time.

Due to the fast development of computer science in the last years, increasingly harder and more complex problems are being faced continuously. A large number of metaheuristics designed for solving such complex problems exists in the literature [3, 5]. Among them, evolutionary algorithms (EAs) are very popular optimization methods [6, 7, 8]. They consist in evolving a population of individuals (tentative solutions), emulating the biological process found in Nature, so that individuals are improved. This family of techniques apply an iterative and stochastic process on a set of individuals (population), where each individual represents a potential solution to the problem. To measure their quality, a fitness value is assigned to each individual. This value represents the quantitative information used by the algorithm to guide the search. The balance between exploration (diversification) of new areas of the search space and exploitation (intensification) of good solutions accomplished by this kind of algorithms is one of the key factors for their high performance with respect to other metaheuristics. This exploration/exploitation tradeoff can be sharpened by tuning some different parameters of the algorithms such as the population used, the variation operators applied, or the probability of applying them, among others.

G. Luque and E. Alba: Parallel Genetic Algorithms, SCI 367, pp. 3–13.
springerlink.com 

In this book we focus on parallel genetic algorithms (pGA), a class of EA in which the tentative solutions are evaluated and evolve in parallel. Although the use of metaheuristics allows to significantly reduce the temporal complexity of the search process, the exploration remains time-consuming for industrial problems. Therefore, parallelism is necessary to not only reduce the resolution time, but also to improve the quality of the provided solutions [6, 9, 10], since in many cases the search progress is conducted differently when using a pGA.

In this chapter we first define some important issues for optimization problems. After that, we present a brief introduction to the field of metaheuristics (Section 1.2) and, particularly to evolutionary algorithms (in Section 1.3). In Section 1.4) we describe the two main existing models of decentralized population in GAs: cellular and distributed GAs since they are often the basis for building parallel GAs. At the end of this chapter, a summary is offered.

## 1.1 Optimization

In this section, we define some basic notations used along this book. Initially, we give a formal definition of (mono-objective) optimization. Assuming minimization (without any lost in generality), we can define an optimization problem as follows:

**Definition 1.1** ***(Optimization).*** *An optimization problem is formalized by a pair (S,f), where $S \neq \emptyset$ represents the solution space -or search space- of the problem, while f is a quality criterion known as the objective function, defined as:*

$$f : S \to \mathbb{R}. \tag{1.1}$$

*Thus, solving an optimization problem consists in finding a set of decision variables values such that the represented solutions by these values $\boldsymbol{i}^* \in S$ satisfy the following inequality:*

$$f(\boldsymbol{i}^*) \leq f(\boldsymbol{i}), \forall \boldsymbol{i} \in S. \tag{1.2}$$

Assuming maximization or minimization does not restrict the generality of the results, we can establish an equivalence between the maximization and minimization problems as:

$$max\{f(\boldsymbol{i})|\boldsymbol{i} \in S\} \equiv min\{-f(\boldsymbol{i})|\boldsymbol{i} \in S\}. \tag{1.3}$$

According to the domain of $S$, we can define *binary optimization* ($S \subseteq \mathbb{B}$), *complete or discrete* ($S \subseteq \mathbb{N}$), *continuous* ($S \subseteq \mathbb{R}$), or *heterogeneous* -or *mixed*- ($S \subseteq \mathbb{B} \cup \mathbb{N} \cup \mathbb{R}$) problems.

A definition of proximity or distance between different solutions of the search space is necessary for solving an optimization problem. Two solutions are close each other if they belong to the same neighborhood in the search space. We define the neighborhood of a solution as:

**Definition 1.2 *(Neighborhood).*** *Being (S,f) an optimization problem, a neighborhood structure in S can be defined as:*

$$N : S \to S, \tag{1.4}$$

*such that for each solution $\boldsymbol{i} \in S$ a set $S_i \subseteq S$ is defined. It also holds that if $\boldsymbol{i}$ is in the neighborhood of $\boldsymbol{j}$, then $\boldsymbol{j}$ is also in the neighborhood of $\boldsymbol{i}$: $\boldsymbol{j} \in S_i$ iff $\boldsymbol{i} \in S_j$.*

In general, in a complex optimization problem, the objective function often presents an optimal solution that is an optimum only in its neighborhood, but which is not optimal if we consider the whole search space. Therefore, a global search method can be easily trapped in an optimal value inside a neighborhood, thus giving rise to the concept of *local optimum*:

**Definition 1.3 *(Local optimum).*** *Being $(S, f)$ an optimization problem, and $S_{i'} \subseteq S$ the neighborhood of a solution $i' \in S_{i'}$, $i'$ is a local optimum if the next inequality is satisfied (assuming minimization):*

$$f(i') \leq f(i), \forall i \in S_{i'}. \tag{1.5}$$

When tackling real life optimization problems, we are usually forced to deal with constraints. In these cases, the area of feasible solutions $S$ is limited to those that satisfy all the constraints. Thus, the next definition is needed:

**Definition 1.4 *(Optimization with constraints).*** *Given an optimization problem $(S, f)$, we define $M = \{i \in S \mid g_k(i) \geq 0, \ \forall k \in [1, \ldots, q]\}$ as the region of feasible solutions of the objective function $f : S \to \mathbb{R}$. The functions $g_k : S \to \mathbb{R}$ are called constraints, and these $g_k$ are named differently according to the value taken by $i \in S$:*

$$\begin{aligned}
&\textit{satisfied} : \Leftrightarrow g_k(i) \geq 0, \\
&\textit{active} \quad : \Leftrightarrow g_k(i) = 0, \\
&\textit{inactive} : \Leftrightarrow g_k(i) > 0, \ \textit{and} \\
&\textit{violated} : \Leftrightarrow g_k(i) < 0,
\end{aligned}$$

*Global optimization problem is the term used when no constraints exist, i.e. iff $M = S$; in other case it is referred to as a restricted or constrained problem.*

## 1.2 Metaheuristics

There exist many proposal of algorithmic techniques in the literature, both exact and approximate, for solving optimization problems (see Fig. 1.1 for a basic taxonomy). Exact algorithms guarantee to find the optimal solution for all the existing (finite set of) instances. Generally, since exact methods need exponential computation times when facing large instances of complex problems, NP-hard problems can not be realistically tackled. Therefore, the use of approximate technique is a rising topic in the last decades. In these methods we lose the guarantee of finding the global optimal solution (often, but not always) in order to find good solutions in a significantly shorter time compared to exact methods. This lost of the guarantee of finding the optimal solution is often practically, since in a number of metaheuristics, the theoretical convergence to the global optimum is proven under some conditions.

For the last twenty years a new kind of approximate techniques has been emerging, consisting basically in combining simple *ad-hoc* heuristic methods (approximate algorithms with stochastic guided components) in higher level search sheets in order to explore and to exploit the search space efficiently and effectively. These methods are commonly known as *metaheuristics*. In [3] the reader can find some metaheuristic definitions given by different authors, but in general we can state that metaheuristics are high level strategies having a given structure that plans the application of a set of operations (variation operators) to explore high dimensional and complex search spaces.

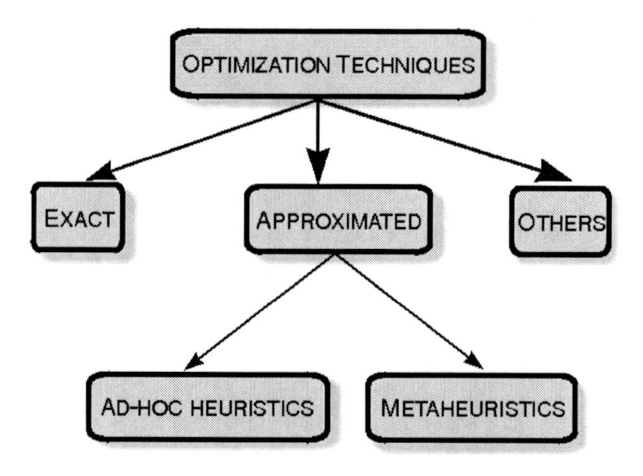


**Fig. 1.1** Classification of optimization problems.

Metaheuristics can be classified in many different ways. In [3] a classification is given according to different properties which characterize them. One of these classifications relies on the number of solutions: population based (work with a set of solutions) and trajectory based (work with a single solution). The latter starts with a single initial solution. At each step of the search the current solution is replaced by another (often the best) solution found in its explored neighborhood. Frequently, such a metaheuristic allows to find a local optimal solution, and so they are called exploitation-oriented methods. On the other hand, population based methods make use of a population of solutions. The initial population is enhanced through a natural (or purely mathematical) evolution process. At each generation of the process, the whole population or a part of the population is replaced by newly generated individuals (usually the best ones). Population based methods are often called exploration-oriented methods. Among the best well-known metaheuristics, we can find evolutionary algorithms (EAs) [8], iterative local search (ILS) [11], simulated annealing (SA) [12], tabu search (TS) [13], variable neighborhood search VNS [14], and ant colony optimization (ACO) [15].

## 1.3 Evolutionary Algorithms

In the seventies and eighties (with various other punctual works before), several researchers coincided in developing, independently of each other, the idea of implementing algorithms based on the organic evolution model in an attempt to solve adaptive and hard optimization tasks on computers. Nowadays, due to their stockiness and large applicability, and also to the availability of higher computational power (e.g., parallelism), the resulting research field, that of evolutionary computation, receives growing attention in all important research agencies in the world.

The evolutionary computation framework [8] stands for a wide set of families of techniques for solving the problem of searching optimal values by using computational models, most of them inspired by evolutionary processes (evolutionary algorithms). Evolutionary Algorithms (EAs) are population based optimization techniques designed for searching optimal values in complex spaces. They are loosely based on some biological processes that can be seen in Nature, like natural selection [16] or genetic inheritance [17] of parental good traits. Part of the evolution is determined by the natural selection of different individuals competing for resources in the environment. Therefore, some individuals are better than others. Those that are better are more likely to survive, learn, and propagate their genetic material.

Sexual reproduction allows some shuffling of chromosomes, producing offspring that contain a combination of information from each parent. This is known as the recombination operation, which is often referred to as crossover because of the way that biologists have observed strands of chromosomes crossing over during the exchange. Recombination happens in an environment where the selection of mates for reproduction is largely a function of the fitness of individuals, i.e., how good each individual is at competing in its environment.

As in the biological case, individuals can occasionally mutate. Mutation is an important source of diversity for EAs. In an EA, a large amount of diversity is usually introduced at the start of the algorithm by randomizing the genes in the population. The importance of mutation, which introduces further diversity while the algorithm is running, is a matter of debate. Some refer to it as a background operator, simply replacing some of the original diversity which may have been lost, while others view it as playing a dominant role in the evolutionary process (e.g., avoiding getting stuck in local optima).

An EA proceeds in an iterative way by successively evolving the current population of individuals. This evolution is usually a consequence of applying stochastic variation operators such as selection, recombination, and mutation on the population in order to compute a whole generation of new individuals. The initial population is usually generated randomly, although it is also normal to use some seeding technique in order to speed up the search by starting from good quality solutions. A fitness evaluation assigns a value to every individual, which is representative of its suitability to the problem at hands. This evaluation can be performed by an objective function (e.g., a mathematical expression or a computer simulation) or by a subjective opinion, in which the best solutions are selected by an external -human- agent (e.g., expert design of furniture or draws using interactive EAs). The stop criterion is usually set to reach a preprogrammed number of iterations of the algorithm, or to find a solution to the problem (or an approximation to it, if it is known beforehand).

Individuals encode tentative solutions to the problem usually in the form of strings (of binary, decimal, or real numbers), trees, and maybe other data structures [18]. Every individual has an assigned fitness value as a measure of its adequacy, so that better fitness values represent better individuals. This fitness value is used for deciding which individuals are better and which ones are worse.

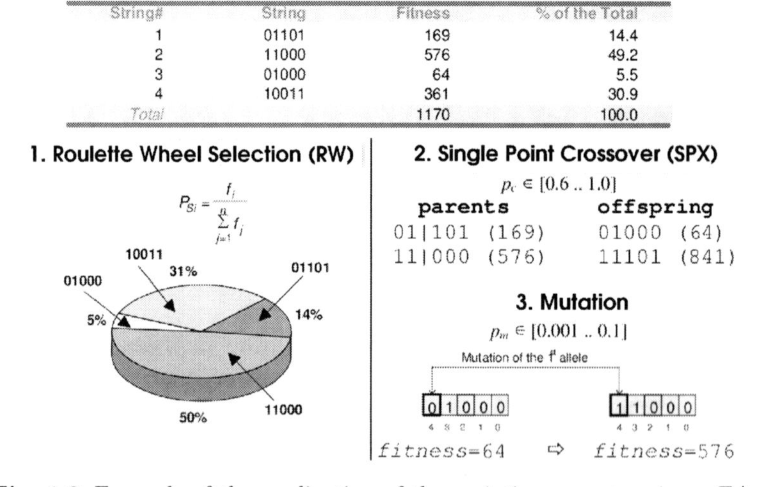

| String# | String | Fitness | % of the Total |
|---|---|---|---|
| 1 | 01101 | 169 | 14.4 |
| 2 | 11000 | 576 | 49.2 |
| 3 | 01000 | 64 | 5.5 |
| 4 | 10011 | 361 | 30.9 |
| Total | | 1170 | 100.0 |



**Fig. 1.2** Example of the application of the variation operators in an EA.

In Fig. 1.2 we show an example of the application of some specific variation operators in a population composed by four individuals. As it can be seen, the "String" column in the upper table is the binary codification of a given problem. The selection operator displayed selects the parents in terms of the percentage of their fitness value with respect to the sum of the fitness values of all the individuals in the population. The recombination operator represented (the single point crossover) splits the chromosomes of the two individuals into two different parts in a randomly chosen location, and then it joints the parts of the different individuals in order to generate two new offspring. Finally, the mutation in this example flips the value of a random gene (the first one in this figure), in order to introduce some more diversity and hopefully getting a better individual, as it is the case of the example.

Now, we analyze in detail the functioning of an evolutionary algorithm. Its pseudo-code is shown in Algorithm 1. As it was said before, evolutionary algorithms work on populations of individuals, which are tentative solutions to the problem. The initial population is usually composed by randomly created individuals, although problem knowledge can help creating faster EAs (e.g., by using a greedy initial feeding of solutions). After the generation of the initial population, the fitness value of each individual is computed, and the algorithm starts off the reproductive cycle. This step lies in generating a new population through the selection of the parents, the recombination of them, the mutation of the offsprings obtained and then, their evaluation. These three variation operators are typical of most EAs, specially GAs, although many EA families usually use less (e.g., evolutionary strategies ES with no recombination) or more (e.g., decentralized EAs) operators. This new population generated in the reproductive cycle ($P'$) will be used, along with the current population ($P$), for obtaining the new population of individuals for the next generation. The algorithm returns the best solution found during the execution.

**Algorithm 1.** Pseudocode for an Evolutionary Algorithm.

```
1: P ← GenerateInitialPopulation()
2: Evaluate(P)
3: while not Termination_Condition() do
4:     P' ← SelectParents(P)
5:     P' ← ApplyVariationOperators(P')
6:     Evaluate(P')
7:     P ← SelectNewPopulation(P,P')
8: end while
9: Return The best solution found;
```

There are two kinds of EAs depending on the replacement used, that is, according to the combination between $P$ and $P'$ for the new generation. Thus, being $\mu$ the number of individuals of $P$ and $\lambda$ the number of individuals in $P'$, if the population of the new generation is obtained from the best $\mu$ individuals of the populations $P$ and $P'$ we have a $(\mu + \lambda)$-EA, meanwhile if the population of the next generation is composed only by the $\mu$ best individuals out of the $\lambda$ belonging to $P'$, we have a $(\mu, \lambda)$-EA. In this second case, it is usual that $\mu \leq \lambda$. The "plus" strategy is inherently elitist (the best solution is always preserved) while the "comma" strategy could led to loosing the best solution from the population.

The application of EAs to optimization (and learning) problems has been very intense during the last decade [8]. In fact, it is possible to find this kind of algorithms applied to solving complex problems like constrained optimization tasks, problems with a noisy objective function, or problems which have high epistasis (high correlation between the values to optimize) and multimodality. The high complexity and applicability of these algorithms has promoted the emergence of innovative new optimization and search models.

Initially, four kinds of evolutionary algorithms [19] could be differentiated. These four families of algorithms were simultaneously developed by different research groups in the world. *Genetic algorithms* (GAs) were initially studied mainly by J. H. Holland [20], in Ann Arbor (Michigan). The *evolution strategies* (ES) were proposed by I. Rechenberg [21] and H.-P. Schwefel [22] in Berlin (Germany), meanwhile the *evolutionary programming* (EP) was firstly proposed by L. J. Fogel [23] in San Diego (California). Last, the fourth family of algorithms, *genetic programming* (GP), appeared two decades later, in 1985, as an adaptation of N. Cramer [24] of a genetic algorithm which worked with tree shaped genes, and it is now widely used thanks to the leading works of J. Koza [25]. In this book, we focus on the GAs since it is maybe the most general (the rest ones are linked to specific encodings or operators) and used algorithms of the four families.

## 1.4 Decentralized Genetic Algorithms

Most GAs use a single population (panmixia) of individuals and apply operators on them as a whole (see Fig. 1.3a). In contrast, a tradition also exists in using structured GAs (where the population is decentralized somehow), especially in relation to their parallel implementation. The use of parallel multiple populations is based on the idea that the isolation of populations allows to keep a higher genetic differentiation [26]. In many cases [27], these algorithms using decentralized populations provide a better sampling of the search space and improve both the numerical behavior and the execution time of an equivalent panmictic algorithm. Among the many types of structured GAs, distributed and cellular algorithms are two popular optimization tools (see Fig. 1.3).

On the one hand, in the case of distributed GAs (dGAs), the population is partitioned in a set of islands in which isolated GAs are executed (demes). Sparse exchanges of individuals are performed among these islands with the goal of introducing some diversity into the subpopulations, thus preventing them from getting stuck in local optima.

On the other hand, in a cellular GA (cGA) the concept of a (small) neighborhood is intensively used; this means that an individual may only interact

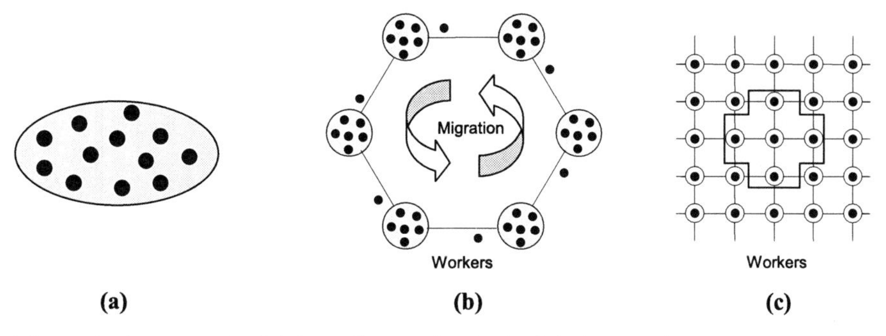


**Fig. 1.3** A panmictic GA has all its individuals (black points) in the same population (a). Structuring the population usually leads to distinguishing between distributed (b) and cellular (c) GAs.

with its nearby neighbors in the breeding loop. The overlapped small neighborhoods of cGAs help in exploring the search space because the induced slow diffusion of solutions through the population provides a kind of exploration (diversification), while exploitation (intensification) takes place inside each neighborhood by genetic operations.

If we think on the population of an GA in terms of graphs, being the individuals the vertices of the graph and the relationship among them the edges, a panmictic GA is a completely connected graph. On the other hand, a cGA is a lattice graph, as one individual can only interact with its nearest neighbors. A dGA is a partition of the panmictic GA into several smaller GAs, that is, in each island we have a completely connected graph (very fast convergence), while there exist only a few connections between the islands.

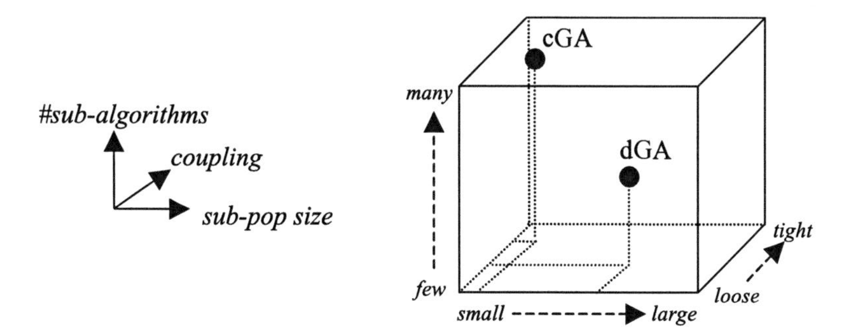


**Fig. 1.4** The structured-population genetic algorithm cube.

These two traditionally decentralized GAs (dGAs and cGAs) are actually two subclasses of the same kind of GA consisting in a set of communicating sub-algorithms. Hence, the actual differences between dGAs and cGAs can be found in the way in which they both structure their populations. In Fig. 1.4, we plot a three-dimensional (3-D) representation of structured algorithms based on the number of subpopulations, the number of individuals in each one, and the degree of interaction among them [28]. As it can be seen, a dGA is composed of several large sub-populations (having >> 1 individuals), loosely connected among them. Conversely, cGAs are made up of a large number of tightly connected sub-populations, each one typically containing only one individual.

This cube can be used to provide a generalized way for classifying structured GAs. However, the points in the cube indicating dGA and cGA are only "centroids"; this means that an arbitrary decentralized algorithm could be hardly classified as belonging to one of two such classes of structured GAs.

In a structured GA, many elementary GAs (grains) exist each, working on separate sub-populations. Each sub-algorithm includes an additional phase of periodic communication with a set of neighboring sub-algorithms located in some topology. This communication usually consists in exchanging a set of individuals or population statistics. All the sub-algorithms were initially studied to perform the same reproductive plan, although there is a recent trend consisting in executing GAs with distinct parameterizations in each subpopulation, thus performing different searches in the space of solutions in each island. This kind of (usually parallel) GAs are called *heterogeneous*.

From this point of view, cellular and distributed GAs only differ in some parameters, and they can be helpful even when run in a monoprocessor. This makes merging them in the same algorithm an interesting option, in order to get a more flexible and efficient algorithm for some kinds of applications [28]. Additionally, any of these GAs (or EAs) can be run in a distributed way (i.e., suited for a workstation cluster or multicore computer), or even on a grid of computers [29] (not to confuse with the "grid" in which individuals evolve in the population of a cellular GA).

## 1.5 Conclusions

In this chapter we have presented evolutionary algorithms, which are iterative techniques operating on a set of individuals composing a population; each of these individuals represents a potential solution to the problem. This population of individuals evolves due to the application of a set of operators inspired in biological processes of Nature, such as natural selection and genetic inheritance. As a result, the individuals of the population are improved during the evolution. GAs are really useful tools for solving complex problems, as they work fast in large (and complex) search spaces thanks to their ability to process multiple solutions simultaneously (concept of population). Hence, EAs can follow different search paths simultaneously, that would be in turn explored in parallel.

Additionally, it is possible to improve the numerical behavior of the algorithm by structuring the population. The main types of GAs with structured populations are distributed and cellular ones. In this book we focus on the distributed case as it was briefly introduced in this chapter. For parallel algorithms, this kind of coarse-grained model of search is exceptionally good and represents a rich set of research lines. We will see in the next chapters that even the panmictic model allows for master-slave parallel execution, thus we will even open our range of study from coarse-grained parallelism to other types of search such as farming and even fine-grained parallelism in modern hardware such as graphic processing units (GPUs) or FPGAs.

# 2

# Parallel Models for Genetic Algorithms

*The whole is more than the sum of the parts.*

*Aristotle (384 - 322 BC) - Greek philosopher*

In the previous chapter we offered a brief introduction to metaheuristics. Now, this chapter is devoted to genetic algorithms (GA) and their parallel models. GAs [4, 20] are stochastic search methods designed for exploring complex problem spaces in order to find optimal solutions, possibly using information of the problem to guide the search. Unlike most other optimization (search, learning) techniques, a population of multiple structures is used by GAs to perform the search along many different areas of the problem space at the same time. The structures composing the population (individuals) encode tentative solutions, which are manipulated competitively by applying them some stochastic operators to find a satisfactory, if not globally, optimal solution.

**Algorithm 2.** Pseudocode of a canonical GA.

```
1: P ← GenerateInitialPopulation()
2: Evaluate(P)
3: while not Termination_Condition() do
4:    P' ← SelectParents(P)
5:    P' ← Recombination(P')
6:    P' ← Mutation(P')
7:    Evaluate(P')
8:    P ← SelectNewPopulation(P,P')
9: end while
10: Return The best solution found;
```

In Algorithm 2, the outline of a classical GA is described. A GA proceeds in an iterative way by successively generating a new population $P(t)$ of individuals from $P(t-1)$, the previous one ($t = 1, 2, 3, \ldots$). The initial population $P(0)$ is generated randomly. A fitness function associates a value to every individual, which is representing its suitability to the problem in hands. The

G. Luque and E. Alba: Parallel Genetic Algorithms, SCI 367, pp. 15–30.
springerlink.com 

canonical algorithm applies stochastic operators such as selection, recombination, and mutation on a population in order to compute a whole generation of new individuals. In a general formulation, variation operators are applied to create a temporary population $P'(t)$, whose individuals are evaluated; then, a new population $P(t+1)$ is obtained by using $P'(t)$ and, optionally, $P(t)$. In GAs, these variations operators are typically recombination and mutation. The stop criterion is usually set as reaching a preprogrammed number of iterations of the algorithm, and/or to find an individual with a given error if the optimum, or an approximation to it, if known beforehand.

For nontrivial problems, the execution of the reproductive cycle of a simple GA may require high computational resources (e.g., large memory and long search times), and thus a variety of algorithmic issues has been studied to design efficient GAs. For this goal, numerous advances are continuously being achieved by designing new operators, hybrid algorithms, termination criteria, and so on [8]. In this chapter, we address one such improvement, consisting in adding parallelism to GAs. In the field of parallel GAs (pGAs) [9], there exists a large number of implementations and algorithms. The reasons of this success have to do first, with the fact that GAs are naturally prone to parallelism, since most variation operators can be easily undertaken in parallel, and second, that using a pGA often takes us not only to use a faster algorithm, but also to get a superior numerical performance [27, 30]. Parallel GAs are characterized by the use of a *structured population* (a spatial distribution of individuals), either in the form of a set of islands [31] or a diffusion grid [32], which is the responsible of such benefits. As a consequence, many authors do not use a parallel machine at all to run structured-population models, and still get better results than with serial traditional GAs [30].

The goal of this chapter is to give a modern classification of the different models and implementations concerning parallel GAs, a field of the evolutionary computation (EC) discipline. In addition, we will empirically test the behavior of some of the most important proposed models, providing to the community what we hope to be a useful baseline comparison for other researchers among the main models for parallelizing GAs existing in the literature.

This chapter is organized as follows. First, a description of the standard model of GA, in which the whole population is considered as a single pool of individuals, is given. In the next section, we address the structured models, in which the population is decentralized somehow. Later, some different implementations of parallel GAs are presented, and a pGA classification is given. In Section 2.4, we test and compare the behavior of several parallel models when solving an instance of the well-known MAXSAT problem. Finally, we summarize our most important conclusions.

## 2.1 Panmictic Genetic Algorithms

In the GA field, it is usual to find algorithms implementing panmictic populations, in which selection takes place globally and any individual can potentially mate with any other one in the population. The same holds for the replacement operator, where any individual can potentially be removed from the pool and replaced by a new one. In contrast, there exists a different (decentralized) selection model, in which individuals are arranged spatially, therefore giving place to *structured GAs* (see Section 2.2). Most other operators, such as recombination or mutation, can be readily applied to these two models.

There exist two popular classes of panmictic GAs, having different granularity at the reproductive step [33]. In the first one, called the "generational" model, a whole new population of $\lambda$ (hence $\lambda = \mu$, $\mu$ being the population size) individuals replaces the old one. The second type is called "steady state", since usually one ($\lambda = 1$) or two ($\lambda = 2$) new individuals are created at every step of the algorithm and then they are inserted back into the population, consequently coexisting with their parents. In Fig. 2.1, we graphically explain these two kinds of panmictic GAs in terms of the number of new individuals being inserted into the population of the next generation ($\lambda$). As it can be seen in the figure, in the mean region (where $1 < \lambda < \mu$), there exists a plethora of selection models generically termed as "generation gap" algorithms, in which a given number of individuals ($\lambda$ value) are replaced with the new ones. Clearly, generational and steady state selection are two special subclasses of generation gap algorithms.

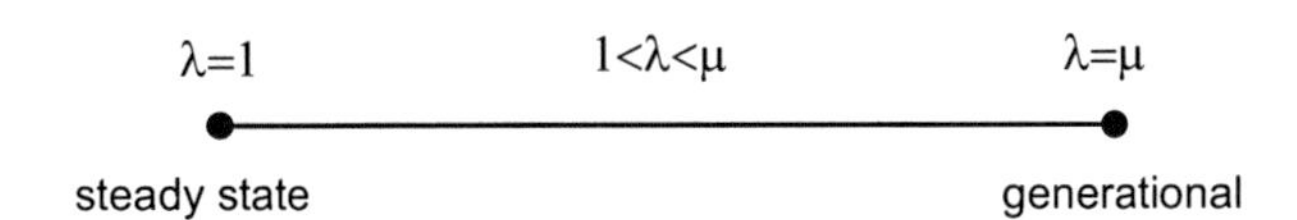


**Fig. 2.1** Panmictic GAs, from steady state to generational algorithms.

Centralized versions of selection are typically found in serial GAs, although some parallel implementations have also used them. For example, the *global parallelism* approach evaluates in parallel the individuals of the population while still using a centralized selection performed sequentially in the main processor running the base algorithm [34]. This algorithm is conceptually the same as the sequential one, although it is faster, especially for time-consuming objective functions. The common drawback of this approach is the bottleneck imposed by the centralized scheme of execution. Usually, the other parts of the algorithm are not worth to be parallelized, unless some population structuring principle is used (see Section 2.2).

Most pGAs found in the literature implement some kind of spatial disposition for the individuals, and then parallelize the resulting chunks in a pool of processors [6]. We must stress at this point of the discussion that parallelization is very often achieved by first structuring the panmictic algorithm, and then parallelizing it. This is why we distinguish throughout this book between structuring populations and making parallel implementations, since the same structured GA can admit many different implementations. The next section is devoted to explain different ways of structuring the populations. The resulting models can be executed in parallel or not, but in fact some structured models suggest a straightforward parallel implementation.

## 2.2 Structured Genetic Algorithms

There exists a long tradition in using structured populations in EC, especially associated to parallel implementations. Among the most widely known types of structured GAs, the *distributed* and *cellular* ones are very popular optimization procedures [35].

Decentralizing a single population can be achieved by partitioning it into several subpopulations, where component GAs are run performing sparse exchanges (migrations) of individuals (dGAs), or in the form of neighborhoods (cGAs). In dGAs, additional parameters controlling when migration occurs and how migrants are selected/incorporated from/to the source/target island are needed [31, 36]. In cGAs, the existence of overlapped small neighborhoods helps in exploring the search space [37]. These two kinds of GAs seem to provide a better sampling of the search space, and to improve the numerical and runtime behavior of the basic algorithm in many cases [27, 30].

The main difference in a cGA with respect to a panmictic GA is its decentralized selection and variation. In cGAs, the reproductive loop is performed inside every one of the numerous individual pools. In a cGA, one given individual has its own pool of potential mates, defined by its neighboring individuals; at the same time, one individual belongs to many pools. This structure with overlapped neighborhoods (usually a 1- or 2-dimensional lattice) is used to provide a smooth diffusion of good solutions across the *grid*. A cGA can be implemented in a distributed memory MIMD computer [38] (e.g., a cluster or multicore), although its more direct implementation is on a SIMD computer (e.g., a FPGA or GPU).

A dGA is a multipopulation (island) model performing sparse exchanges of individuals among the elementary populations. This model can be readily implemented in distributed memory MIMD computers, which provides one main reason for its popularity. A migration policy controls the kind of dGA being used. The migration policy must define the island topology, when migration occurs, which individuals are being exchanged, the synchronization among the subpopulations, and the kind of integration of exchanged individuals within the target subpopulations. The advantages of a distributed model (either running on separate processors or not) is that it is usually faster (it reduces the required numerical steps) than a panmictic GA. The reason for this is that the run time and the number of evaluations are potentially reduced thanks to its separate search in several regions from the problem space. A high diversity and species formation are two of their well-reported features [39].

So far, we have made the implicit hypothesis that the genetic material, as well as the evolutionary conditions, such as selection and recombination methods, were the same for all the individuals and all the populations of a structured GA. Let us call these algorithm types *uniform* or *homogeneous*. If one gives up some of these constraints and allows different subpopulations to evolve with different parameters and/or with different individual representations for the same problem, then new distributed algorithms may arise. We will name these algorithms *nonuniform* or *heterogeneous* parallel GAs. Tanese did some original work in this field and was the first in studying the use of different mutation and crossover rates in different populations [39]. A more recent example of nonuniform algorithms is the *injection island GA* (iiGA) of Lin *et al.* [40]. In an iiGA, there are multiple populations that encode the same problem using a different representation size, and thus different resolutions in different islands. The migration rules are also special in the sense that migration is only one-way, going from a low- to a high-resolution node. According to Lin *et al.*, such a hierarchy has a number of advantages with respect to a standard island algorithm. A similar hierarchical topology approach has been recently used in [41] with some differences such as real-coded GAs and two-way migration. The purported advantages are: no need for representation conversion, better precision, and better exploration of the search space using a nonuniform mutation scheme.

A related proposal has been offered by Herrera *et al.* [42]. Their *gradual distributed real-coded GA* involves a hierarchical structure in which a higher level nonuniform distributed GA joins a number of uniform distributed GAs that are connected among themselves. The uniform distributed GAs differ in their exploration and exploitation properties due to different crossover methods and selection pressures. The proposed topology is the cube-connected cycle in which the lower level distributed GAs are rings at the corners of the cube, and the rings are connected at the higher level along the edges of the cube. There are two types of migration: local migration among subpopulations in the same lower level distributed GA and global migrations between subpopulations belonging to different lower level distributed GAs. According to [43], the proposed scheme outperforms other distributed GAs on the set of test functions that were used in the paper. A deeper explanation with details and a summary on research using heterogeneous pGAs can be found in [44, 45].

## 2.3 Parallel Genetic Algorithms

In this section, our goal is to present a structured vision of the parallel models and parallel implementations of GAs. Therefore, Subsection 2.3.1 is devoted to describe the parallel model used to parallelize a GA, Subsection 2.3.2 presents a classification of the parallel implementations, and Subsection 2.3.3 focuses on some of the most promising research lines in the field of pGAs.

### *2.3.1 Parallel Models*

This subsection briefly describes the primary conceptual models of the major parallel GA paradigms that are implemented in the literature.

#### Independent Runs Model

Many researchers use a pool of processors to speed up the execution of a sequential algorithm, just because *independent runs* can be made more rapidly by using several processors than by using a single one. In this case, no interaction at all exists among the independent runs. This extremely simple method of doing simultaneous work can be very useful. For example, it can be used to run several versions of the same problem with different initial conditions, thus allowing gathering statistics on the problem in a short time. Since GAs are stochastic in nature, it is very important the availability of this kind of statistics.

#### Master-Slave Model

The Master-Slave model is easy to visualize. It consists in distributing the objective function evaluations among several slave processors while the main loop of the GA is executed in a master processor. This parallel paradigm is quite simple to implement and its search space exploration is conceptually identical to that of a GA executing on a serial processor. In other words, the number of processors being used is independent of which solutions are evaluated, except for time. This paradigm is illustrated in Fig. 2.2, where the master processor sends parameters (those necessary for the objective function evaluation) to the slaves; objective function values are then returned when computed.

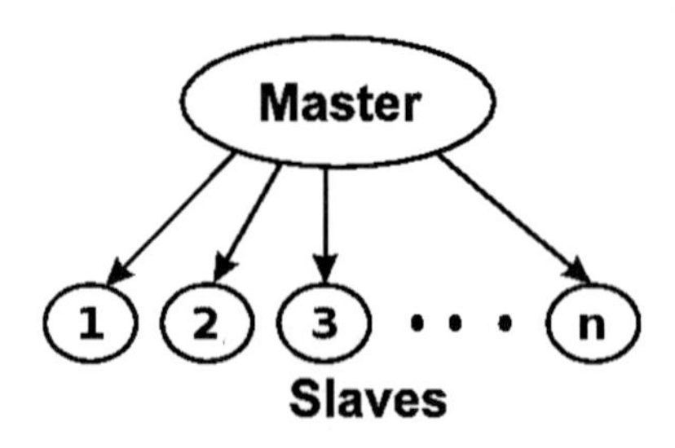


**Fig. 2.2** Master-Slave parallel model.

The master processor controls the parallelization of the objective function evaluation tasks (and possibly the fitness assignment and/or transformation) performed by the slaves. This model is generally more efficient as the objective evaluation becomes more expensive to compute, since the communication overhead is insignificant with respect to the fitness evaluation time. However, it tends to present a bottleneck for some dozens processors, thus usually preventing scalability.

### Distributed Model

In distributed GAs, the population is structured into smaller subpopulations relatively isolated one from the others. Parallel GAs based on this paradigm are sometimes called multi-population or multi-deme GAs. Regardless their name, the key characteristic of this kind of algorithm is that (a copy of) individuals within a particular subpopulation (or island) can occasionally migrate to another one. This paradigm is illustrated in Fig. 2.3. Note that the communication channels shown are logical; specific assignments are made as a part of the GA's migration strategy and are mapped to some physical network.

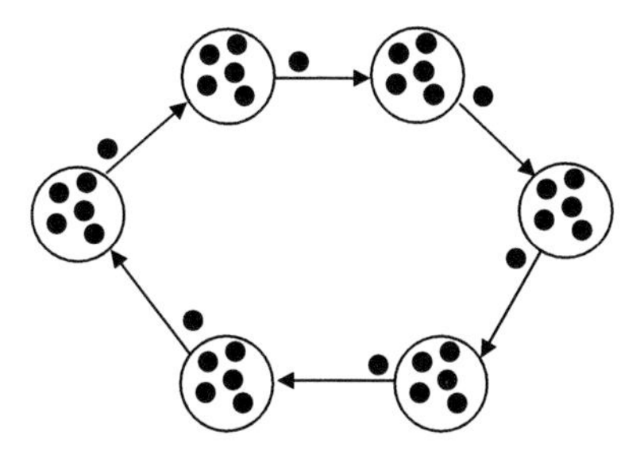

**Fig. 2.3** Distributed model, also called multiple-deme, multipopulation, or coarse-grained parallel model.

Conceptually, the overall GA population is partitioned into a number of independent, separate subpopulations (or demes). An alternative view is that of several small, separate GAs executing simultaneously. Individuals occasionally migrate between one particular island and its neighbors, although these islands usually evolve in isolation for the majority of the GA runtime. Here, genetic operators (selection, mutation, and recombination) take place within each island, which means that each island can search in very different regions of the whole search space with respect to the others. As said before, each island could also have different parameter values (heterogeneous GAs [45]). The distributed model requires the identification of a suitable migration policy. The main parameters of the migration policy include the following ones:

- *Migration Gap.* Since a dGA usually makes sparse exchanges of individuals among the subpopulations, we must define the migration gap, this is the number of steps in every subpopulation between two successive exchanges (steps of isolated evolution). It can be activated in every subpopulation either periodically or by using a given probability $P_M$ to decide in every step whether migration will take place or not.

- *Migration Rate.* This parameter determines the number of individuals that undergo migration in every exchange. Its value can be given as a percentage of the population size or else as an absolute value.
- *Selection/Replacement of Migrants.* This parameter decides how to select emigrant solutions, and which solutions have to be replaced by the immigrants. It is very common in parallel distributed GAs to use the same selection/replacement operators for dealing with migrants.
- *Topology.* This parameter defines the neighbor of each island, i.e., the islands that a concrete subpopulation can send to (or receive from) individuals. The traditional nomenclature divides parallel GAs into *stepping-stone* and *island* models, depending on whether individuals can freely migrate to any subpopulation or if they are restricted to migrate to geographically nearby islands, respectively.

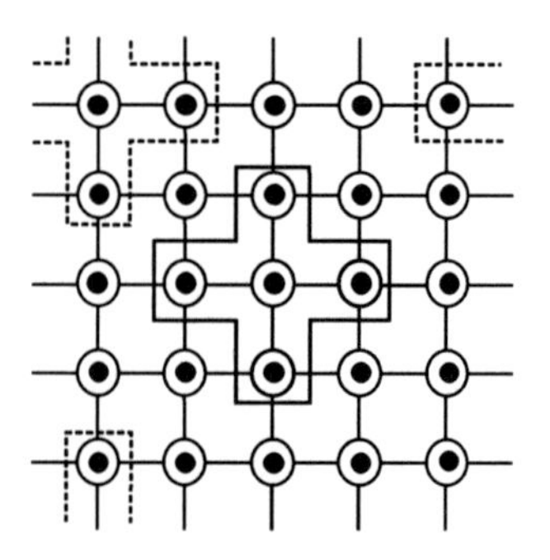

**Fig. 2.4** Cellular model, also called fine-grained parallel model.

### Cellular Model

Like the master-slave model, the parallel cellular (or diffusion) GA paradigm normally deals with a single conceptual population, where each processing unit holds just a few individuals (usually one or two). That is the reason because this model is sometimes called *fine-grained* parallelism. The main characteristic of this model is the structuring of the population into neighborhoods, where individuals may only interact with their neighbors. Thus, since good solutions (possibly) arise in different areas of the overall topology, they are slowly spread (or diffused) throughout the whole structure (population). This model is illustrated in Fig. 2.4.

Cellular GAs were initially designed for working in massively parallel machines, although the model itself has been adopted also for distributed systems [38] and monoprocessor machines [30, 46]. This issue may be stated clearly, since many researchers still hold in their minds the relationship between massively parallel GAs and cellular GAs, what nowadays this is mostly not true.

Today's graphical processing units (GPUs) and field programmable gate array (FPGA) represent new hardware support for actually running cGAs on massively parallel platforms [47, 48]

### Other Models

It is also common to find many implementations of difficult classification in the literature. In general, they are called *hybrid parallel algorithms* since they implement characteristics of different parallel models.

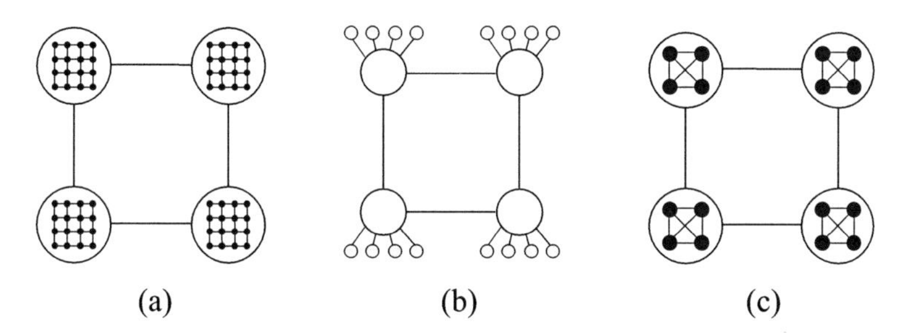


**Fig. 2.5** Hybrid parallel models: (a) cellular distributed GA, (b) distributed master-slave GA, and (c) two level distributed GA.

For example, Fig. 2.5 shows three hybrid search architectures in which a two-level approach of parallelization is undertaken. In the three cases the highest level of parallelization is a dGA. In Fig. 2.5a, the basic islands perform a cGA, thus trying to get the combined advantages of the two models. In Fig. 2.5b, we have many global parallelization farms connected in a distributed fashion, thus exploiting parallelism for making fast evolutions and for obtaining separate population evolutions at the same time. Finally, Fig. 2.5c presents several farms of distributed algorithms with a still higher level of distribution, allowing migration among connected farms. Although these combinations may give rise to interesting and efficient new algorithms, they have the drawback of needing some additional new parameters to account for a more complex topology structure.

## *2.3.2 A Brief Survey on Parallel GAs*

In this section we briefly discuss the main features of some of the most important pGAs by presenting a structured classification, organized by the model of parallelization (other classifications can be found in [6]).

In Table 2.1, we provide a quick overview of different pGAs to point out important milestones in parallel computing with GAs. These "implementations" have rarely been studied as "parallel models". Instead, usually only the implementation itself is evaluated.

**Table 2.1** A quick survey of several parallel GAs.

| Algorithm | Article | Parallel Model |
|---|---|---|
| ASPARAGOS | [49] (1989) | Fine grain. Applies hill-climbing if no improvement. |
| dGA | [31] (1989) | Distributed populations. |
| GENITOR II | [50] (1990) | Coarse grain. |
| ECO-GA | [51] (1991) | Fine grain. |
| PGA | [52] (1991) | Sub-populations, migrate the best, local Hill-Climbing. |
| SGA-Cube | [53] (1991) | Coarse grain. Implemented on the nCUBE 2. |
| EnGENEer | [54] (1992) | Global parallelization. |
| PARAGENESIS | [55] (1993) | Coarse grain. Made for the CM-200. |
| GAME | [55] (1993) | Object oriented set of general programming tools. |
| PEGAsuS | [56] (1993) | Coarse or fine grain. High-level programming on MIMD. |
| DGENESIS | [57] (1994) | Coarse grain with migration among sub-populations. |
| GAMAS | [58] (1994) | Coarse grain. Uses 4 species of strings (nodes). |
| iiGA | [40] (1994) | Injection island GA, heterogeneous and asynchronous. |
| PGAPack | [34] (1995) | Global parallelization (parallel evaluations). |
| CoPDEB | [59] (1996) | Coarse grain. Every subpop. applies different operators. |
| GALOPPS | [60] (1996) | Coarse grain. |
| MARS | [61] (1999) | Parallel environment with fault tolerance. |
| RPL2 | [62] (1999) | Coarse grain. Very flexible to define new GA models. |
| GDGA | [43] (2000) | Coarse grain. Hypercube topology. |
| DREAM | [63] (2002) | Framework for distributed EAs. |
| Hy4 | [45] (2004) | Coarse grain. Heterogeneous and hypercube topology. |
| MALLBA | [64] (2004) | An efficient general framework for parallel algorithms. |
| ParadisEO | [65] (2004) | A general framework for parallel algorithms. |
| JGAP | [66] (2008) | A framework for genetic algorithms and genetic programming. |
| apcGA | [67] (2009) | Fine grain. |

Some coarse-grain algorithms like dGA [31], DGENESIS [57], GALOPPS [60], PARAGENESIS [55], and PGA [52] are relatively close to the general model of migration islands. They often include many features to improve efficiency. Some other coarse-grain models like CoPDEB [59], GDGA [43], and Hy4 [45] have been designed for specific goals, such as providing explicit exploration/exploitation by applying different operators on each island. Another example of this class is the iiGA [40], which promotes coding and operator heterogeneity (see Section 2.2). A further parallel environment providing adaptation with respect to the dynamic behavior of the computer pool and fault tolerance is MARS, described by Talbi *et al.* in [61].

Some other pGAs execute nonorthodox models of coarse-grain evolution. This is the case of GAMAS [58], based on using different alphabets in every island, and GENITOR II [50], based on a steady-state reproduction.

In contrast, fine grain pGAs have been strongly associated to the massively parallel machines on which they run: ASPARAGOS [49] and ECO-GA [51]; although recently, it is also find some implementations of fine grain algorithms for distributed platforms [67]. This is also the case of models of difficult classification like PEGAsuS [56], or SGA-Cube [53]. Depending on the global parallelization model, some implementations, such as EnGENEer [54] or PGAPack [34], are available.

Finally, it is worth to emphasize some efforts to construct general frameworks for PEAs, like GAME [55], PEGAsuS, RPL2 [62], DREAM [63], MALLBA [64], and ParadisEO [65]. These systems are endowed with "general" programming structures intended to ease the implementation of any model of PEA for the user, who must particularize these general structures to define his/her own algorithm. Nowadays, many researchers are using object-oriented programming (OOP) to create a higher quality software for pGAs, but unfortunately some of the most important issues typical in OOP are continuously being ignored in the resulting implementations. The reader can find some general guidelines for designing object-oriented PEAs in [68].

All these models and implementations offer different levels of flexibility, ranging from a single pGA to the specification of general pGA models. This list is not complete, of course, but it helps in understanding the current "state of the art".

### *2.3.3 New Trends in pGAs*

In this section, we focus on some of the most promising research lines in the field of pGAs. Future achievements should take note of these issues.

- *Tackling dynamic function optimization problems (DOP):* pGAs will have an important role in optimizing complex functions whose optima vary in time (learning-like process). Such problems consist in optimizing a successive set of (logically different) fitness functions, each one usually being a (high/small) perturbation of the precedent one. Industrial processes, like real task-scheduling, and daily life tasks such as controlling an elevator or the traffic light system can be dealt with dynamic models. Some pGAs, like cGAs and dGAs, can face such DOP environments successfully thanks to their natural diversity enhancements and speciation-like features [69, 70].
- *Developing theoretical issues:* Improving the formal explanations on the influence of parameters on the convergence and search of pGAs will endow the research community with tools allowing to analyze, understand, and customize a GA family for a given problem [71, 72, 73].
- *Running pGAs on geographically separated clusters:* This will allow the user to utilize sparsely located computational resources in a metacomputing fashion in order to solve his/her optimization problem. A distinguished example of such a system is to use the Web as a pool of processors to run pGAs for solving the same problem. In particular, Grid computing [29] and Peer-to-Peer (P2P) computing [74] have become a real alternative to traditional supercomputing for the development of parallel applications that harness massive computational resources. This is a great challenge, since nowadays grid and P2P-enabled frameworks for metaheuristics are just emerging [45, 63, 65].

- *New parallel platforms:* Exploiting new hardware architectures like GPUs and multicore computers by developing new specialized pGAs profiting from these architectures to run more efficiently [47, 48, 75, 76, 77].
- *Benchmarking soft computing techniques:* At present, it is clear that a widely available, large, and standard set of problems is needed to assess the quality of existing and new pGAs. Problem instances of different difficulty, specially targeted to test the behavior of pGAs and related techniques can greatly help practitioners in choosing the most suitable pGA or hybrid algorithm for the task in hands.

## 2.4 First Experimental Results

In this section, we perform several experimental tests to study the behavior of the different parallel models described in the previous sections. Concretely, we use a parallel distributed GA (**dGA**), a parallel cellular GA (**cGA**), a parallel master-slave GA (**MS**), and a parallel distributed GA where cooperation among the islands does not exist, i.e., the islands are completely independent (**idGA**, isolated dGA). All these models are implemented on a distributed-memory system.

Let us give some details about the implementation of the parallel **cGA** on distributed systems (the rest admit a straightforward implementation on a computer cluster). In this algorithm the whole population is divided among the processors, but the global behavior of this parallel cGA is the same as that of a sequential (cellular) one. At the beginning of each iteration, all the processors send the individuals of their first/last column/row to their neighbor islands (see Fig. 2.6). After receiving the individuals from the neighbors, a sequential cGA is executed in each subpopulation. The remaining of the algorithms has a canonical implementation and no especial issues are used.

For testing the parallel algorithms we have used the well-known MAXSAT problem. The next subsection briefly describes the MAXSAT problem and then we discuss the results.

### *2.4.1 MAXSAT Problem*

The satisfiability (SAT) problem is commonly recognized as a fundamental problem in artificial intelligence applications, automated reasoning, mathematical logic, and related fields. The MAXSAT is a variant of this general problem.

Formally, the SAT problem can be formulated as follows. Let $U = \{u_1, \ldots, u_n\}$ be a set of $n$ Boolean variables. A truth assignment for $U$ is a function $t : U \rightarrow \{true, false\}$. Two literals, $u$ and $\neg u$, can match with each variable. A literal $u$ (resp. $\neg u$) is true under $t$ if and only if $t(u) = true$ (resp. $t(\neg u) = false$). A set $C$ of literals is called a clause and it represents the disjunction (*or* logical connective). A set of clauses is called a formula. A

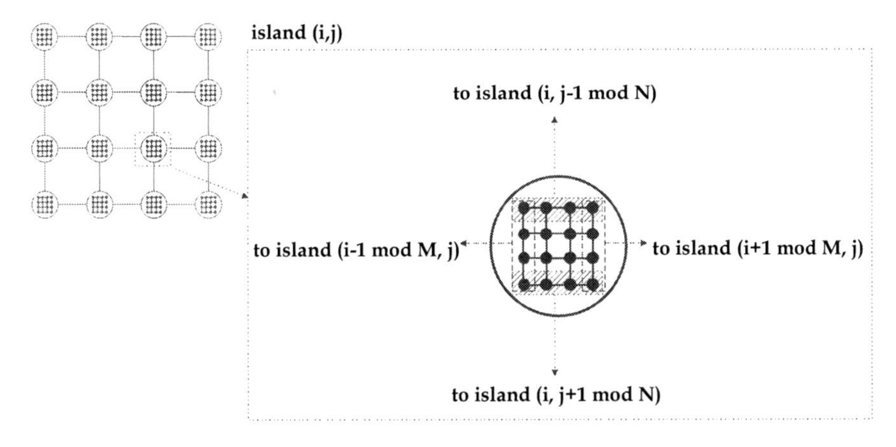


**Fig. 2.6** Parallel implementation of the distribution of the cGA (M is the number of columns and N is the number of rows.)

formula $f$ is interpreted as a formula of the propositional calculus in conjunctive normal form (CNF) so that a truth assignment $t$ satisfies a clause $C$ iff at least one literal $u \in C$ is true under $t$. Finally, $t$ satisfies $f$ iff it satisfies every clause in $f$. The SAT problem consists of a set of $n$ variables $\{u_1, \ldots, u_n\}$ and a set of $m$ clauses $C_1, \ldots, C_m$. The goal is to determine whether or not there exists an assignment of truth values to variables that makes the formula $f = C_1 \wedge \cdots \wedge C_m$ in CNF satisfiable. Among the extensions to SAT, MAXSAT [78] is the most well-known one. In this case, a parameter $K$ is given and the problem is to determine whether there exists an assignment $t$ of truth values to variables such that at least $K$ clauses are satisfied. SAT can be considered as a special case of MAXSAT when $K$ equals the number $m$ of clauses.

In the experiments we use the first instance of De Jong [79]. This instance is composed of 100 variables and 430 clauses ($f^*(optimum) = 430$).

## 2.4.2 Analysis of Results

In this section, we study the behavior of different parallel implementations of a GA when solving the MAXSAT problem. We begin with a description of the parameters of each algorithm. No special configuration analysis has been made for determining the optimum parameter values for each algorithm. The whole population is composed of 800 individuals. In the evaluated parallel implementations, each processor has a population of $800/n$, where $n$ is the number of processors. All the algorithms use the one-point crossover operator (with probability 0.7) and bit-flip mutation operator (with probability 0.2). In distributed GAs, the migration occurs in a unidirectional ring manner, sending one single randomly chosen individual to the neighbor subpopulation. The target population incorporates this individual only if it is better than its current worst solution. The migration step is performed every 20 iterations in every island in an asynchronous way. All the experiments are performed on 16 Pentium 4 at 2.8 GHz PCs linked by a Fast-Ethernet communication network. Because of the stochastic nature of genetic algorithms, we perform 100 independent runs of each test to gain sufficient experimental data.

Now, let us begin the analysis by presenting in Table 2.2 the number of executions that found the optimal value (**% hit** column), the number of needed evaluations (**# evals** column) and the running time in seconds (**time** column) for all the algorithms: the sequential GA, the parallel master-slave GA, the parallel isolated distributed GA, the (cooperative) parallel distributed GA, and the parallel cellular GA (the parallel algorithms running on 2, 4, 8 and 16 processors). We also show in Table 2.3 the speedup of these algorithms. We use the *Weak* definition of speedup [6], i.e., we compare the parallel implementation runtime with respect to the serial one (see Chapter 3 for an in depth discussion on speedup).

**Table 2.2** Average Results for all the parallel GAs.

| Alg. | % hit | # evals | time | Alg. | % hit | # evals | time |
|---|---|---|---|---|---|---|---|
| **Seq.** | 60% | 97671 | 19.12 | | | | |
| **MS2** | 61% | 95832 | 16.63 | **idGA2** | 41% | 92133 | 9.46 |
| **MS4** | 60% | 101821 | 14.17 | **idGA4** | 20% | 89730 | 5.17 |
| **MS8** | 58% | 99124 | 13.64 | **idGA8** | 7% | 91264 | 2.49 |
| **MS16** | 62% | 96875 | 12.15 | **idGA16** | 0% | - | - |
| **dGA2** | 72% | 86133 | 9.87 | **cGA2** | **85%** | 92286 | 10.40 |
| **dGA4** | 73% | 88200 | 5.22 | **cGA4** | 83% | 94187 | 5.79 |
| **dGA8** | 72% | **85993** | 2.58 | **cGA8** | 83% | 92488 | 2.94 |
| **dGA16** | 68% | 93180 | **1.30** | **cGA16** | 84% | 91280 | 1.64 |

If we interpret the results in Table 2.2 we can notice several facts. First, let us analyze each parallel model. As we expected, the behavior of the master-slave algorithms (**MSx**) are similar to the sequential version since they obtain the same number of hits and sample a similar number of points of the search space (they are statistically equivalent). As expected, the master-slave methods spend a lower time to find the optimum but the actual gain is very low (see Table 2.3) for any number of processors. This happens because the execution time of the fitness function does not compensate the overhead of the communications for these problem instances.

The **idGA** model allows to reduce the search time and obtains a very good speedup (see also Table 2.3), nearby linear, but the results are worse than those of the serial algorithms (lower number of hits), and even with 16 processors it can not find the optimal solution in any execution. This is not surprising, since increasing the number of processors means decreasing the population size, and the algorithm is not able to maintain enough diversity to find the global solution.

**Table 2.3** Weak Speedup.

| | Speedup | | | |
|---|---|---|---|---|
| **Alg.** | $n = 2$ | $n = 4$ | $n = 8$ | $n = 16$ |
| **MS*n*** | 1.14 | 1.34 | 1.40 | 1.57 |
| **idGA*n*** | 2.02 | 3.69 | 7.67 | - |
| **dGA*n*** | 1.93 | 3.66 | 7.41 | 14.7 |
| **cGA*n*** | 1.83 | 3.30 | 6.50 | 11.65 |

A clear conclusion is that the distributed GAs are better than the sequential algorithm both numerically and in terms of search time, since they obtain a higher number of hits with a lower number of evaluations, and they also reduce the total search time. The speedup is quite good but it is always sublinear and it slightly moves away from the linear speedup when the number of CPUs increases. That is, as the number of CPUs increases a small loss of efficiency is obtained.

Numerically speaking, the cellular GAs are the best ones, since they obtain the highest number of hits. Surprisingly, they also show very low execution times, which are only slightly worse than those of the dGAs, which perform a meaningful lower number of migration exchanges. This is worth of further research since this means that numerical and real time efficiency are improved at the same time what is a noticeable result (see [67] for more details).

## 2.5 Summary

This chapter contains a modern survey of parallel models and implementations of GAs. By summarizing the parallel algorithms, their applications, classes, and theoretical foundations we intend to offer valuable information not only for beginners, but also for researchers working with GAs or heuristics in general.

As we have seen along this chapter, the robustness and advantages of sequential GAs are enhanced when pGAs are used. The drawback is the more complex analysis and design, and also the need of some kind of parallel platform to run it.

A classified overview on the most important up-to-date pGA systems is discussed. In this chapter, not only a survey of existing problems is outlined, but also possible variants (apart from the basic operations) and future research trends are considered, yielding what we hope is a unified overview and a useful text. The cites in this chapter have been elaborated to serve as a directory for granting the reader access to the valuable results that parallel GAs are offering to the research community.

Finally, we have performed an experimental test with the most common parallel models used in the literature. We have used distributed, cellular, master-slave, and independent runs models to solve the well-known

MAXSAT. For this problem, we noticed that the master-slave model is not suitable since the overhead provoked by the communications is not compensated by the reductions in the execution time of the objective function. The isolated dGA (or independent runs model) obtains very low execution times, but the solution quality gets worse. The use of distributed and cellular models managed to improve both the number of hits and the search time. The distributed version spends shorter time than the cellular one (i.e., dGA is more efficient) since it performs a low number of exchanges, but the cellular GA obtains the best number of hits of all the studied algorithms (i.e., cGA is more accurate).

# 3

# Best Practices in Reporting Results with Parallel Genetic Algorithms

*Disdain for rules is as harmful as their excessive observation.*

*Juan Luis Vives (1492 - 1540) - Spanish philosopher*

Most optimization tasks found in real world applications impose several constraints that frequently prevent the utilization of exact methods. The complexity of these problems (they are often NP-hard [78]) or the limited computational resources available to solve them (time, memory) have made the development of metaheuristics a major field in present research. In these cases, metaheuristics provide optimal or suboptimal feasible solutions in a reasonable time. Although the use of metaheuristics allows to significantly reduce the time of the search process, the high dimension of many tasks will always pose problems and result in time-consuming scenarios for industrial problems. Therefore, parallelism is an approach not only to reduce the resolution time, but also to improve the quality of the provided solutions. This last holds since parallel algorithms usually run a different search model with respect to sequential ones [35] (see also Chapter 1).

Unlike exact methods, where time-efficiency is a main measure for evaluating their success, there are two chief issues in evaluating parallel metaheuristics: how fast solutions can be obtained, and how far they are from the optimum. We can distinguish between two different approaches for analyzing metaheuristics: a theoretical analysis (worst-case analysis, average-case analysis, ...) and an experimental analysis. Several authors [80, 81] have developed theoretical analyses of some importance for a number of heuristics and problems. But, these theoretical achievements have a difficulty that makes it hard their utilization for most realistic problems and algorithms, severely limiting their range of application. As a consequence, most of metaheuristics are evaluated *empirically* in an *ad-hoc* manner.

An experimental analysis usually consists in applying the developed algorithms to a collection of problem instances and comparatively report the observed solution quality and consumed computational resources (usually time). Other researchers [82, 83] have tried to offer a kind of methodological framework to deal with the experimental evaluation of heuristics. Each a

G. Luque and E. Alba: Parallel Genetic Algorithms, SCI 367, pp. 31–51.
springerlink.com

methodological approach mainly motivates this chapter. Important aspects of an evaluation are the experimental design, finding good sources of test instances, measuring the algorithmic performance in a meaningful way, sound analysis, and clear presentation of results. Due to the great difficulty of making all this correctly, the actual main issues of the experimental evaluation are simplified to just *highlight* some guidelines for designing experiments, and reporting on their results. An excellent algorithmic survey about simulations and statistical analysis is given in [84]. In that paper, McGeoch includes an extensive set of basic references on statistical methods and a general guide for designing experiments.

In this chapter, we focus on how the experiments should be performed, and how the results must be reported in order to make fair comparisons between parallel metaheuristics in general, and parallel genetic algorithms. Specially, we are interesting in revising, proposing, and applying parallel performance metrics and statistical analysis guidelines to ensure that our conclusions are correct.

This chapter is organized as follows. The next section briefly summarizes some parallel metrics such as speedup and related performance measures. Section 3.2 discusses how to report results on parallel metaheuristics. Then, in a later section, we perform several practical experiments to illustrate the importance of the metric in the achieved conclusions. Finally, some concluding remarks are outlined in Section 3.5.

## 3.1 Parallel Performance Measures

There are different metrics to measure the performance of parallel algorithms. In the first subsection we discuss in detail the most common measure, i.e., the speedup, and address its meaningful utilization in parallel metaheuristics. Later, in a second subsection, we summarize some other metrics also found in the literature.

### *3.1.1 Speedup*

The most important measure of a parallel algorithm is the *speedup* [82, 85]. This metric computes the ratio between sequential and parallel times. Therefore, the definition of time is the first aspect that we must face. In a uniprocessor system, a common performance measure is the *CPU time* to solve the problem; this is the time the processor spends executing algorithm instructions, typically excluding the time for input of problem data, output of results, and system overhead activities. In the parallel case, time is not a sum of CPU times on each processor, neither the largest among them. Since the objective of parallelism is the reduction of the real-time, time should definitely include any overhead activity because it is the price of using a parallel algorithm. Hence the most prudent choice for measuring the performance of a parallel code is the *wall-clock time* to solve the problem at hands. This means using the time between starting and finishing the whole algorithm. For a fair comparison, then this should also be the time for the sequential case.

The speedup compares the serial time against the parallel time until a given stopping condition is fulfilled. If we denote by $T_m$ the execution time for an algorithm using $m$ processor, the standard speedup is the ratio between the *faster* execution time on a uni-processor $T_1$ and the *actual* execution time on $m$ processors $T_m$:

$$s_m = \frac{T_1}{T_m} \tag{3.1}$$

For non-deterministic algorithms we cannot use this metric directly. For this kind of methods, the speedup should instead compare the *mean* serial execution time against the *mean* parallel execution time:

$$s_m = \frac{E[T_1]}{E[T_m]} \tag{3.2}$$

Here we are assuming a normal distribution of time or at least that we have many independent runs to approximate it (central limit theorem). Otherwise, the median or other parameter could be also used.

With this definition we can distinguish among: *sublinear* speedup ($s_m < m$), *linear* speedup ($s_m = m$), and *superlinear* speedup ($s_m > m$). The main difficulty with that measure is that researchers do not agree on the meaning of $T_1$ and $T_m$. In his study, Alba [85] distinguishes between several definitions of speedup depending of the meaning of these values (see Table 3.1).

**Table 3.1** Taxonomy of speedup measures proposed by Alba [85].

| |
|---|
| I. Strong speedup |
| II. Weak speedup |
|     A. Speedup with solution stop |
|         1. Versus panmixia |
|         2. Orthodox |
|     B. Speed with predefined effort |

*Strong speedup* (type I) compares the parallel run time against the best-so-far sequential algorithm. This is the most exact and standard definition of speedup, but due to the difficulty of finding the current most efficient algorithm, most designers of parallel algorithms do not use it. *Weak speedup* (type II) compares the parallel algorithm developed by a researcher against his/her own serial version. In this case, two stopping criterion for the algorithms exist: solution *quality* or maximum *effort*. The author discards the latter because it is against the aim of speedup to compare algorithms not yielding results of equal accuracy. In fact, after the standard and historical use of speedup, the two compared algorithms should be making exactly the same search work, some thing that it is clearly not interesting in most metaheuristics.

He proposes two variants of the weak speed with solution stop: to compare the parallel algorithm against the canonical sequential version (type II.A.1) or to compare the run time of the parallel algorithm on one processor against the run time of the same algorithm on $m$ processors (type II.A.2). In the first case we are comparing two clearly different algorithms what again could raise serious concerns on the meaning of this metric. Thus, the orthodox approach (type II.A.2) seems the fairest way of evaluating parallel metaheuristics and at the same time using a standard metric found in other parallel applications.

Barr and Hickman in [82] showed a different taxonomy: *Speedup*, *Relative speedup* and *Absolute speedup*. The *Speedup* measures the ratio between the time of the faster serial code on a parallel machine with the time of the parallel code using $m$ processors on a machine with similar characteristics to the one used in the serial one. The *Relative speedup* is the ratio of the serial execution time with parallel code on one processor with respect to the execution time of that code on $m$ processors. This definition is similar to the type II.A.2 shown above. The *Absolute speedup* compares the fastest serial time on any computer with the parallel time on $m$ processors. This metric is the same as the strong speedup defined by [85].

As a conclusion, it is clear that the evaluated parallel metaheuristics should compute solutions having a similar accuracy as the sequential ones. This accuracy could be the optimal fitness value (if known) or a relaxation of it (e.g., 90%), but in any case the same value. Just in this case we are allowed to compare times. The used times are average mean times: the parallel code on one machine versus the parallel code on $m$ machines. All this define a sound way for comparisons, both practical (no best algorithm needed) and orthodox (same codes, same accuracy).

### Superlinear Speedup

Although several authors have reported superlinear speedup [36, 40], its existence is always controversial. Anyway, based on past experiences, we can expect to get superlinear speedup sometimes whatever the parallel metric program or algorithm is. In fact, we can point out several sources behind superlinear speedup:

- *Implementation source:* The algorithm being run on one processor is "inefficient" in some way. For example, if the algorithm uses lists of data, the parallel one can be faster because it manages shorter lists (or trees, with a faster access). On the other hand, the parallelism can simplify several operations of the algorithm.
- *Numerical source:* Since the search space is usually very large, the sequential program may have to search a large portion before finding the required solution. On the other hand, the parallel version may find the

solution more quickly due to the change in the order in which the space is searched.

- *Physical source:* When moving from a sequential to a parallel machine, it is often the case that one gets more than just an increase in CPU power (see Fig. 3.1). Other resources, such as memory, caché, etc. may also increase linearly with the number of processors. A parallel metaheuristic may achieve superlinear speedup by taking advantage of these additional resources (purposely or not). That is, in a parallel platform, each subpopulation is a different machine and it can fit into cachés while in a sequential execution, all the subpopulations are in the same machine, and then we have to use the main memory to store them since they do not fit in the caché.

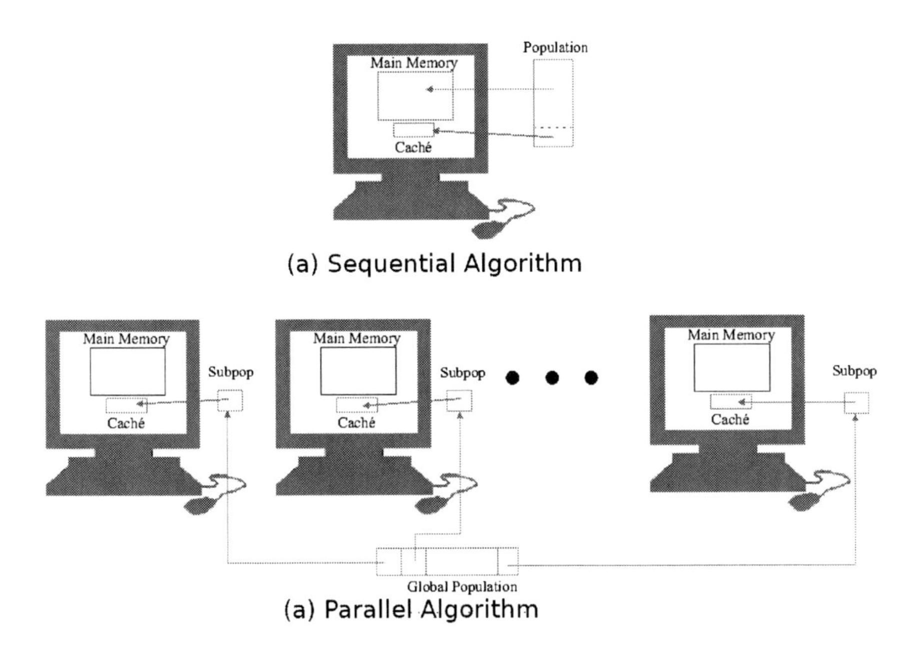


**Fig. 3.1** Physical source for superlinear speedup. The population (a) does not fit into a single caché, but when run in parallel (b), the resulting chunks do fit, providing superlinear values of speedup.

We therefore conclude that superlinear speedup is possible theoretically and as a result of empirical tests in literature, both for homogeneous [86] and heterogeneous [87, 88] computing networks.

### 3.1.2 Other Parallel Measures

Although the speedup is a widely used metric, there exist other measures of the performance of a parallel GA or general metaheuristic.

The *efficiency* (Equation 3.3) is a normalization of the speedup and allows to compare different algorithms ($e_m = 1$ means linear speedup).

$$e_m = \frac{s_m}{m} \tag{3.3}$$

There exist several variants of the efficiency metric. For example, the *incremental efficiency* (Equation 3.4) shows the fraction of time improvement from adding another processor; it is also very used when the uni-processor times are unknown. This metric has been later generalized (Equation 3.5) to measure the improvement attained by increasing the number of processors from $n$ to $m$.

$$ie_m = \frac{(m-1) \cdot E[T_{m-1}]}{m \cdot E[T_m]} \tag{3.4}$$

$$gie_{n,m} = \frac{n \cdot E[T_n]}{m \cdot E[T_m]} \tag{3.5}$$

The previous metrics indicate the improvement coming from using additional processing elements, but they do not measure the utilization of the available memory. The *scaled speedup* (Equation 3.6) addresses this issue, and allows to measure the full utilization of hardware resources:

$$ss_m = \frac{Estimated\ time\ to\ solve\ problem\ of\ size\ n \cdot m\ on\ 1\ processor}{Actual\ time\ to\ solve\ problem\ of\ size\ n \cdot m\ on\ m\ processors} \tag{3.6}$$

where $n$ is the size of the largest problem which may be stored in the memory associated to one processor. Its major disadvantage is that performing an accurate estimation of the serial time is difficult and it is impractical for many problems.

Closely related to *scaled speedup* is *scaleup*, but is not based on an estimation uni-processor time:

$$su_{m,n} = \frac{Time\ to\ solve\ k\ problems\ on\ m\ processors}{Time\ to\ solve\ n \cdot k\ problems\ on\ n \cdot m\ processors} \tag{3.7}$$

This metric measures the ability of the algorithm to solve a $n$-times larger job on a $n$-times larger system in the same time as the original system. Therefore, linear speedup occurs when $su_{m,n} = 1$.

Finally, Karp and Flatt [89] have devised an additional interesting metric for measuring the performance of any parallel algorithm that can help us to identify much more subtle effects than using speedup alone. They call it the *serial fraction* of the algorithm (Equation 3.8).

$$f_m = \frac{1/s_m - 1/m}{1 - 1/m} \tag{3.8}$$

Ideally, the serial fraction should stay constant for an algorithm when using different values of $m$ (number of processors). If the absolute speedup value is small in our experimental study we can still say that the result is good if $f_m$ remains constant for different values of $m$, since the loss of efficiency is due to the limited parallelism of the program. On the other side, smoothly increments of $f_m$ with $m$ is a warning that the granularity of the parallel tasks is too fine. A third scenario is possible, in which a significant reduction in $f_m$ occurs with growing values of $m$, indicating something akin to superlinear speedup. If superlinear speedup occurs, then $f_m$ would take a negative value.

## 3.2 How to Report Results in pGAs

In general, the goal of a scientific publication is to present a new approach or algorithm that works better, in some sense, than existing algorithms. This requires experimental tests to compare the new algorithm with respect to the rest. It is, in general, hard to make fair comparisons between algorithms. The reason is that we need to ensure the same experimental settlement (computer, network protocols, operating system, ...), and, assuming this is correctly done, that later can infer different conclusions from the same results depending on the metrics we use. This is specially important for non-deterministic methods. In this section we address the main issues on experimental testing for reporting numerical effort results, and the statistical analysis that *must* be performed to ensure that the conclusions are meaningful. The main steps are shown in Fig. 3.2.

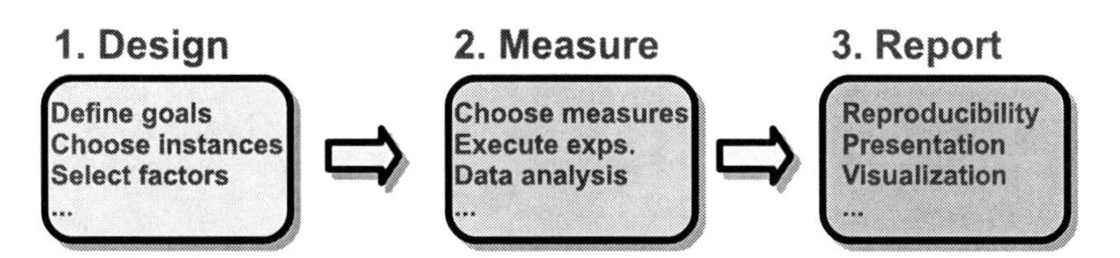


**Fig. 3.2** Main steps for an experimental design.

### 3.2.1 *Experimentation*

The first choice that a researcher must make is the problem domain and the problem instances to test his/her algorithm. That decision depends on the goals of the experimentation. We can distinguish between two clearly different objectives: (1) optimization and (2) understanding of the algorithms.

Optimizing is a commonly practiced sport in designing a metaheuristic that beats others on a given problem or set of problems. This kind of experimental research finishes by establishing the superiority of a given heuristic over others. In this scenario, researchers should not be limited to establishing *that* one metaheuristic is better than another in some way, but also to investigate *why*. A very good study of this latter subject can be found for example in [90].

One important decision is the instance used. The set of instances must be complex enough to obtain interesting results and must have a sufficient variety of scenarios to allow some generalization of the conclusions. Problem generators [79] are specially good for a varied and wide analysis. In the next paragraphs we show the main classes of instances (a more comprehensive classification can be found in [83, 91]).

### Real World Instances

The instances taken from real applications represent a hard testbed for testing algorithms. Sadly, it is rarely possible to obtain more than a few real data for any computational experiment due to proprietary or economic considerations. An alternative is to use random variants of real instances, i.e., the structure of the problem class is preserved, but details are randomly changed to produce new instances.

Another approach is using *natural instances* [91], that represent instances that emerge from a specific real life situation, such as timetabling of a school. This class of instances has the advantage of being freely available. Specially, academic instances must be analyzed in the existing literature to not reinvent the wheel, and to avoid using straightforward benchmarks [92].

### Standard Instances

In this class are included the instances, benchmarks, and problem instance generators that, due to their wide use in experimentation, became standard in the specialized literature. For example, Reinelt [93] offers the *TSPLIB*, a travelling salesman problem test instances, Demirkol *et al.* [94] offer something similar for job scheduling problems. Such libraries allow to test specific issues of algorithms and also to compare our results against other methods, or CEC and GECCO instances for continuous global optimization [95, 96, 97]. The OR-library [98] is a final excellent example of results (academy plus industry) for a large set of problem classes.

### Random Instances

Finally, when none of the mentioned sources provide an adequate supply for tests, the remaining alternative is pure random generation. This method is the faster way to obtain a diverse group of test instances, but is also the most controversial.

After having selected a problem or a group of instances, we must design the computational experiments. Generally, the design starts by analyzing the effects of several factors on the algorithm performance. These factors include problem factors, such as problem size, number of constrains, etc., plus algorithmic factors, such as parameters or components used for the search of the optimum. If the cost of the computer experiments are low, we can do a *full factorial* design, but in general, it is not possible due to the large number of experiments: we usually need to reduce the factors. There is a wide literature on *fractional factorial* design in statistics, which seeks to assess the same effects of a fractional analysis without running all the combinations of influencing parameters (see for example [99]). RACE [100] and SPO [101] is also an interesting approach for validating and reducing the effort of the researcher in experimentation.

The next step in an experimental project is to execute the experiments, choose the measure of performance, and analyze the data. These steps are addressed in the next sections.

### *3.2.2 Measuring Performance*

Once that we have chosen the instances that we are going to use, and the factors that we are going to analyze, we must select the appropriate measures for the goal of our study.

The objective of a metaheuristic is to find a *good* solution in a *reasonable* time. Therefore, the choice of performance measures for experiments with heuristics necessarily involves both solution quality and computational effort. Because of the stochastic nature of metaheuristics, a number of independent experiments need to be conducted to gain sufficient experimental data. The performance measures for these heuristics are based on some kind of statistics.

### *3.2.3 Quality of the Solutions*

This is one of the most important issues to evaluate the performance of an algorithm. For instances where the optimal solution is known, one can easily define a measure: the success rate or *number of hits*. This measure can be defined as the percentage of runs terminating with success (**% hits**). But this metric cannot be used in all cases. For example, there are problems where the optimal solution is not known at all and a lower/upper bound is also unavailable. In other cases, although the optimum is known, its calculation delays too much, and the requirements must be relaxed down to find a *good* approximation in a reasonable time. It is also a common practice in metaheuristics for the experiments to have a specific bound of computational effort (a given number of search space points visited or a maximum execution time).

In all these cases, when optimum is not known or located, statistical metrics are also used. Most popular metrics include the mean and the median of the best performing solutions, such as the fitness (a measure of the quality of the solution) over all executions. These values can be calculated for any problem. For each run of a given metaheuristic the best fitness can be defined as the fitness of the best solution at termination. For parallel metaheuristics it is defined as the best global solution found by the set of cooperating algorithms.

In a problem where the optimum is known, nothing prevents us to use both **% hits** and median/mean of the final quality (or of the effort). Furthermore, all combinations of low/high values can occur for these measures. We can obtain a low number of hits and a high mean/median accuracy; this, for example, indicates a robust method, that seldom achieves the optimal solution. An opposite combination is also possible but it is not common. In that case, the algorithm achieves the optimum in several runs but the rest of the runs compute a very bad fitness.

In practice, a simple comparison between two averages or medians might not give the same result as a comparison between two statistical distributions. In general, it is necessary to offer additional statistical values such as the variance, and to perform a global statistical analysis to ensure that the conclusions are meaningful and not just random noise. We discuss this issue in Subsection 3.2.5.

### *3.2.4 Computational Effort*

While algorithms that produce more accurate solutions are important, the speed of their computation is a key factor. Within metaheuristics, the computational effort is typically measured by the number of evaluations of the objective function and/or the execution time. In general, the number of evaluations is defined in terms of the number of points of the search space visited.

Many researchers prefer the number of evaluations as a way to measure the computational effort, since it eliminates the effects of particular implementations, software, and hardware, thus making comparisons independent from such details. But this measure can be misleading in several cases in the field of parallel methods. For example, if some evaluations take longer than others (for example, in parallel genetic programming [25]) or if an evaluation can be done very fast, then the number of evaluations does not reflect the algorithm's speed correctly. In some cases the concept of evaluation could even not exists (such in ACO) or blur (like in GRASP or TS). Also, the traditional goal of parallelism is not the reduction of the number of evaluations but the reduction of time. Therefore, a researcher should usually report the two metrics to measure the computational effort.

It is very common to use the average evaluations/time to a solution, defined over those runs that end in a solution (with a predefined quality maybe different from the optimal one). Sometimes the average evaluations/time to termination is used instead of the average evaluations/time to a solution of a given accuracy. This practice has clear disadvantages, i.e., for runs finding solutions of different accuracy, using the total execution time/effort to compare algorithms becomes hard to interpret from the point of view of the parallelism. On the contrary, imposing a predefined time/effort and then compare the solution quality of the algorithms is an interesting and correct metric; what it is incorrect is to use in this same case also the run times to compare algorithms, i.e., to measure speedup of efficiency (although works making this can be found in literature). The reason is that imposing a number of evaluations directly determines the search time and the speedup itself.

### 3.2.5 Statistical Analysis

In most papers, the objective is to prove that a particular heuristic outperforms another one. But as we said before, the comparison between two average values might be different from the comparison between two distributions. Therefore, statistical methods should be employed wherever possible to indicate the strength of the relations between different factors and performance measures.

Usually, the researchers use $t$-test or an analysis of variance (ANOVA) to ensure the *statistical significance* of the results, i.e., determining whether an observed effect is likely to be due to sampling errors. Several statistical methods and the conditions to apply them are shown in Fig. 3.3 [10]. Firstly, we should decide between non-parametric and parametric tests; when the data are non-normal and there are not many experimental data (number of experiments $< 30$) should use non-parametric methods otherwise parametric test can be used. Kolmogorov-Smirnov test is a powerful, accurate and low-cost method to check data normality. The Student $t$-test is widely used to compare means of normal data. This method can be only used when there are two populations. In other case, we must use ANOVA test and a later analysis to compare and sort means. For non-normal data, a wide set of methods have been proposed (see Fig. 3.3). All of these methods assume several hypotheses to obtain a proper behavior, e.g., they assume a linear relation between causes and effects.

The $t$-test is based on the *Student's t* distribution. It allows to calculate the statistical significance of two samplings with a given confidence level, typically between 95% ($p$-value $< 0.05$) and 99% ($p$-value $< 0.01$). The underlying notion of ANOVA is to assume that every non-random variation in experimental observations is due to differences in mean performance at alternative levels of the experimental factors. ANOVA proceeds by estimating each of the various means and partitioning the total *sum of squares*, or squared deviation from the sample global mean into separate parts due to each experimental factor and to error.

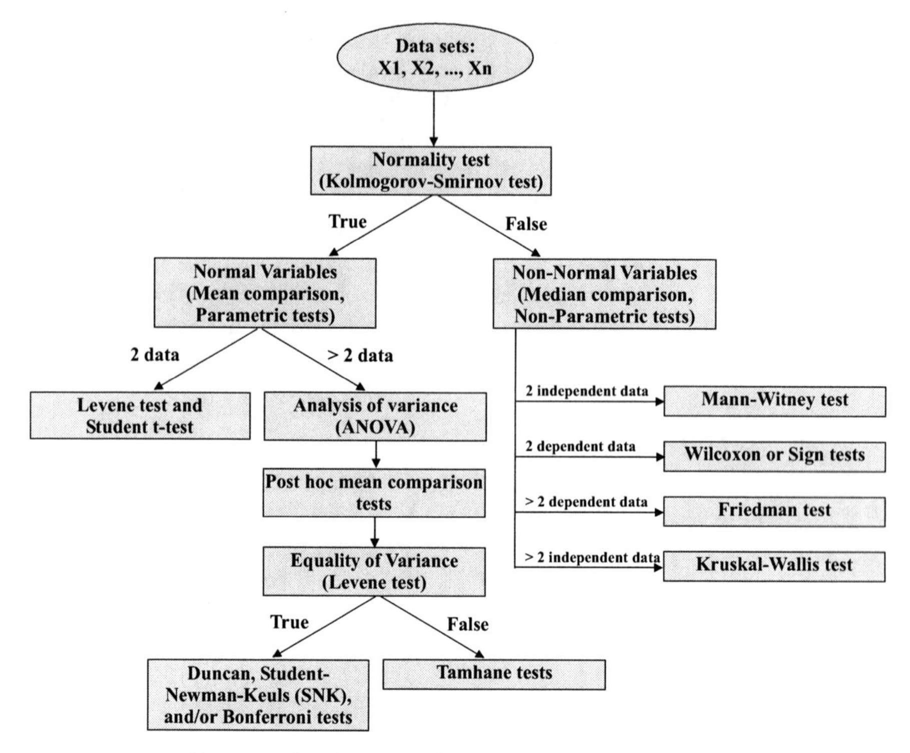


**Fig. 3.3** Application scheme of statistical methods.

The two analyses, $t$-test and ANOVA, can only be applied if the source distribution is normal. In metaheuristics, the resulting distribution could easily be non-normal. For this case, there is a theorem that helps. The *Central Limit Theorem* states that the sum of many identically distributed random variable tends to a Gaussian. So the mean of any set of samples tends to a normal distribution. But in several cases the Central Limit Theorem is not useful and too general. In these cases, there are a host of nonparametric techniques (for example, the sign test) that can and should be employed to sustain the author's arguments, even if the results show no statistical difference between the quality of the solutions produced by the metaheuristics under study [102].

### *3.2.6 Reporting Results*

The final step in an experimental design is to document the experimental details and findings, and to communicate them to the international community. In the next paragraphs, we show the most important issues that should be taken into account.

### Reproducibility

One necessary part of every presentation should be the background on how the experiment was conducted. The reproducibility is an essential part of scientific research, and experimental results that cannot be independently verified are given little credit in the scientific community. Hence, the algorithm and its implementation should be described in sufficient detail to allow replication, including any parameter (probabilities, constants, ...), problem encoding, pseudo-random number generation, etc. The source and characteristics of problem instances should also be documented. Besides, many computing environment factors, that can influence the empirical performance of a method, should be documented: number, types and speeds of processors, size and configuration of memories, communication network, operating system, etc. We advice always including summary tables of algorithms parameters and computational resources in every scientific problem.

### Presenting Results

A final important detail is the presentation of the results. The best way to support your conclusion is to display your data in such a way as to highlight the trends it exhibits, the distinctions it makes, and so forth. There are many good display techniques depending on the types of points one wants to make (for example, see [103] or [104]).

Tables by themselves are usually a very inefficient way of showing the results. Hence, if there is any graphical way to summarize the data and reveal its message, it is almost to be preferred to a table alone. On the other hand, although pictures can often tell your story more quickly, they are usually a poor way of presenting the details of your results. Therefore, a scientific paper should contain both pictures and tables.

## 3.3 Inadequate Utilization of Parallel Metrics

The objective of a parallel metaheuristic is to find a *good* solution in a *short* time. Therefore, the choice of performance measures for these algorithms necessarily involves both solution quality and computational effort. In this section, we discuss some scenarios, where these metrics are incorrectly used, and we propose a solution to these situations.

1. **Computational effort evaluation:** Many researchers prefer the number of evaluations as a way to measure the computational effort since it eliminates the effects of particular implementations, software, and hardware, thus making comparisons independent from such details. But this measure can be misleading in several cases in the field of parallel methods. Whenever the standard deviation of the average fitness computation is high, for example, if some evaluations take longer than others (parallel

genetic programming [25]) or if the evaluation time is not constant, then the number of evaluations does not reflect the algorithm's speed correctly. Also, the traditional goal of parallelism is not only the reduction of the number of evaluations but the reduction of time. Therefore, a researcher must often use the two metrics to measure this effort.
2. **Comparing means vs. comparing medians:** In practice, a simple comparison between two averages or medians (of time, of solutions quality, ...) might not give the same result as a comparison between two statistical distributions. In general, it is necessary to offer additional statistical values such as the variance, and to perform a global statistical analysis to ensure that the conclusions are meaningful and not just random noise. The main steps are the following: first, a normality test (e.g., Kolmogorov-Smirnov) should be performed in order to check whether the variables follow a normal distribution or not. If so, an Student $t$-test (two set of data) or ANOVA test (two or more set of data) should be done, otherwise we should perform a non parametric test such as Kruskal-Wallis. Therefore, the calculation of speedup is only adequate when the execution times of the algorithms are *statistically different*. This two-step procedure also allows to control the I type error (the probability of incorrectly rejecting the null hypothesis when it is true), since the two phases are independent (they test for different null hypotheses).
3. **Comparing algorithms with different accuracy:** Although this scenario can be interesting in some specific applications, the calculation of the speedup or a related metric is not correct because it is against the aim of the speedup to compare algorithms not yielding results of equal accuracy, since the two algorithms are actually solving two different problems (i.e., it is nonsense, e.g., to compare a sequential metaheuristic solving TSP of 100 cities against its parallel version solving a TSP of 50 cities). Solving two different problems is what we actually have in speedup if the final accuracy is different in sequential and parallel.
4. **Comparing parallel versions vs. canonical serial one:** Several works compare the canonical sequential version of an algorithm (e.g., a panmictic GA) against a parallel version. But these algorithms have a different behavior and therefore, we are comparing clearly different methods (as meaningless such as using the search times of a sequential SA versus a parallel GA in the speedup equation).
5. **Using a predefined effort:** Imposing a predefined time/effort and then comparing the solution quality of the algorithms is an interesting and correct metric in general; what it is incorrect is to use it to measure speedup or efficiency (although works making this can be found in literature). On the contrary, these metrics can be used when we compare the average time to a given solution, defined over those runs that end in a solution (with a predefined quality maybe different from the optimal one). Sometimes, the average evaluations/time to termination is used instead of the average evaluations/time to a solution of a given accuracy. This practice has clear

disadvantages, i.e., for runs finding solutions of different accuracy, using the total execution effort to compare algorithms becomes hard to interpret from the point of view of parallelism.

## 3.4 Illustrating the Influence of Measures

In this section we perform several experimental tests to show the importance of the selected metrics in the conclusions. We use several parallel genetic algorithms and one parallel simulated annealing to solve the well-known MAXSAT problem. Before beginning with the examples, we do a brief background of the algorithms, the problem and the configuration for reproducibility.

### The Algorithms

In the experiments, we use three different parallel models of GA: independent runs (IR), distributed GA (dGA), and a cellular GA (cGA). In the first model, a pool of processors is used to speed up the execution of separate copies of a sequential algorithm, just because *independent runs* can be made more rapidly by using several processors than by using a single one. In dGAs [31], the population is structured into smaller subpopulations relatively isolated from the others. The key feature of this kind of algorithm is that individuals within a particular subpopulation (or island) can occasionally migrate to another one in an asynchronous manner. The parallel cGA [32] paradigm normally deals with a single conceptual population, where each processor holds just a few individuals. The main characteristic of this model is the structuring of the population into neighborhood structures, where individuals may only interact with their neighbors.

Also, we consider a local search method such as simulated annealing. A simulated annealing (SA) [12] is a stochastic technique that can be seen as a hill-climber with an internal mechanism to escape from local optima. For this, moves that increase the energy function being minimized are accepted with a decreasing probability. In our parallel SA there exist multiple asynchronous component SAs. Each component SA periodically exchanges the best solution found (cooperation phase) with its neighbor SA in the ring.

### The Problem

The satisfiability (SAT) problem is commonly recognized as a fundamental problem in Computer Science. The MAXSAT [78] is a variant of this general problem. This problem can be formulated as follows: given a formula $f$ of the propositional calculus in conjunctive normal form (CNF) with $m$ clauses and $n$ variables, the goal of this problem is to determine whether or not there exists an assignment $t$ of truth values to variables such that all clauses are satisfied. In the experiments we use several instances generated by De Jong

[79] (a detailed explanation of this problem can be found in the previous chapter). These instances are composed of $n = 100$ variables and $m = 430$ clauses ($f^*(optimum) = 430$).

### Configuration

No special analysis has been made for determining the optimum parameter values for each algorithm. We use a simple representation for this problem: a binary string of length $n$ (the number of variables) where each digit corresponds to a variable. A value of 1 means that its associated variable is *true*, and 0 defines the associated variable as *false*.

In our GA methods, the whole population is composed of 800 individuals and each processor has a population of $800/m$ individuals, where $m$ is the number of processors. All the GAs use the one-point crossover operator (with probability 0.7) and bit-flip mutation operator (with probability 0.2). In distributed GAs, the migration occurs in a unidirectional ring manner, sending one single randomly chosen individual to the neighbor subpopulation. The target population incorporates this individual only if it is better than its current worst solution. The migration step is performed every 20 iterations in every island in an asynchronous way. For the SA method, we use a proportional update of the temperature, and the cooling factor is set to 0.9. The cooperation phase is performed every 10,000 evaluations. All experiments are performed on Pentium 4 at 2.8 GHz linked by a Gigabit Ethernet communication network. We performed 100 independent runs of each experiment to ensure statistical significance.

In the next subsection we present several examples of utilization of the performance measures. We will highlight how wrong selections of metrics affect to the achieved conclusions.

## *3.4.1 Example 1: On the Absence of Information*

We begin our test showing the results of a SA with different number of processors to solve an instance of MAXSAT. The results can be seen in Table 3.2. The values showed are the number of executions that found the optimal value (**% hit** column), the fitness of the best solution (**best** column), the average fitness (**avg** column), the number of evaluations (**# evals** column) and the running time (**time** column).

In this example, the algorithms did not find an optimal solution in any execution. Then, we cannot use the percentage of hits to compare them: we must stick to a different metric to compare the quality of solutions. We could use the best solution found, but that single value does not represent the actual behavior of the algorithms since the methods are non-deterministics. In this case the best measure to compare the quality of results is the average fitness. Then, we can conclude that the SA with 8 processors is better than the same one with 16 processors. But before starting such conclusion we need to perform a statistical test to ensure the significance of this claim.

**Table 3.2** Results of Example 1.

| Alg. | % hit | best | avg | # evals | time |
|---|---|---|---|---|---|
| **SA8** | 0 % | 426 | 418.3 | - | - |
| **SA16** | 0 % | 428 | 416.1 | - | - |

To assess the statistical significance of the results we performed 100 independent runs (30 independent runs is usually thought as a minimum in metaheuristics). Also, we computed a Student $t$-test analysis so that we could be able to distinguish meaningful differences in the average values. The significance $p$-value is assumed to be 0.05, in order to indicate a 95% confidence level in the results.

In this case, the resulting $p$-value is 0.143, i.e., there is *not* a significant difference among the results. This comes as no surprise, since they are the same algorithm, and the behavior should be similar, while only the time should have been affected by the change in the number of processors. Thus, if we had stated on the superiority of one of them, we would have been mistaken. This clearly illustrated the importance of computing the $p$-value.

In this example, we can not measure the computational effort, since they do not achieve an optimum. In the next example, we will use a different stopping criterion to allow us to compare the computational effort.

### *3.4.2 Example 2: Relaxing the Optimum Helps*

Again, we show in Table 3.3 the results of the same SA as in the previous subsection, but for this test we consider as an optimum any solution with $f^*(x) > 420$ (the global optimum has a fitness value $= 430$).

**Table 3.3** Results of Example 2.

| Alg. | % hit | best | avg | # evals | time |
|---|---|---|---|---|---|
| **SA8** | 60 % | 426 | 418.3 | 60154 | 2.01 |
| **SA16** | 58 % | 428 | 416.1 | 67123 | 1.06 |

For this example, we do not compare the quality of the solution since there is not statistical difference, and therefore we focus on the computational effort. The algorithm with 8 processors, SA8, performs a slightly smaller number of evaluations than the SA16, but the difference is not significant (the $p$-value is larger than 0.05). On the other hand, the reduction in the execution time is significant ($p$-value $= 1.4e^{-5}$). Thus we could have stated at a first glance that SA8 is numerically more efficient than SA16, but statistics tell us that no significant improvement can be drawn. However, we can state that SA16 is better from a time efficiency point of view than SA8. These results are somewhat expected: the behavior (quality of solutions and number of evaluations) of both methods are similar, but the execution time is reduced when the number of processors is increased. Concluding on SA8's numerical superiority should have been a mistake that can be avoided thanks to the utilization of such statistical tests.

### 3.4.3 Example 3: Clear Conclusions do Exist

Some authors think that you can never get conclusions. We will here that in controlled experiments you actually can. Now, let us compare two different algorithms: a parallel GA using the independent run model, and the SA of the previous examples. Both, GA and SA are distributed on 16 machines. As it occurs in the first example, none of the two methods achieve the optimum solution in any independent run. Therefore, we consider the optimum definition of the second example (fitness $> 420$).

**Table 3.4** Results of Example 3.

| **Alg.** | % hit | best | avg | # evals | time |
|---|---|---|---|---|---|
| **IR16** | 37 % | 424 | 408.3 | 85724 | 1.53 |
| **SA16** | 58 % | 428 | 416.1 | 67123 | 1.06 |

Table 3.4 shows a summary of the results for this experiment. From this table we can infer that SA16 is better in all aspects (solution quality, number of evaluations, and time) than IR16. And, this time, these conclusions are all supported by statistical tests, i.e., their $p$-values are all smaller than 0.05. Although SA16 is better than IR16, none of them are adequate for this problem, since they two are quite far from the optimum solution, but this is a different story.

### 3.4.4 Example 4: Meaningfulness Does Not Mean Clear Superiority

Now, we compare the results obtained with the same parallel GA (independent runs model) using two, four, and eight processors. The overall results of this example are shown in Table 3.5.

The statistical test are always positive, i.e., all results are significantly different from the other ones. Then we can conclude that the IR paradigm allows to reduce the search time and obtains a very good speedup, nearby linear, but its results are worse than those of the serial algorithms, since its percentage of hits is lower. It might be surprising, since the algorithm is the same in all the cases, and the expected behavior should be similar. The reason is that, as we increase the number of processors, the population size decreases and the algorithm is not able to keep diversity enough to find the global solution.

**Table 3.5** Results of Example 4.

| **Alg.** | % hit | best | avg | # evals | time | speedup |
|---|---|---|---|---|---|---|
| **Seq.** | 60% | 430 | 419.8 | 97671 | 19.12 | - |
| **IR2** | 41% | 430 | 417.7 | 92133 | 9.46 | 1.98 |
| **IR4** | 20% | 430 | 412.2 | 89730 | 5.17 | 3.43 |
| **IR8** | 7% | 430 | 410.5 | 91264 | 2.49 | 7.61 |

### *3.4.5 Example 5: Speedup: Avoid Comparing Apples against Oranges*

In this case, we show an example on speedup. In Table 3.6 we show the results for a sequential GA against a distributed cellular GA with different number of processors. In this example we focus on the **time** column (seconds). The ANOVA test for this column (sequential vs. parallel) is always significant ($p$-value $= 0.0092$).

**Table 3.6** Result of Example 5.

| **Alg.** | % hit | best | avg | # evals | time |
|---|---|---|---|---|---|
| **Seq.** | 60% | 430 | 421.4 | 97671 | 19.12 |
| **cGA2** | 85% | 430 | 427.4 | 92286 | 10.40 |
| **cGA4** | 83% | 430 | 426.7 | 94187 | 5.79 |
| **cGA8** | 83% | 430 | 427.1 | 92488 | 2.94 |
| **cGA16** | 84% | 430 | 427.0 | 91280 | 1.64 |

As we do not know the best algorithm to this MAXSAT instance, we cannot use the *strong speedup* (see Table 3.1). Then, we must use the weak definition of speedup. On the data of Table 3.6, we can measure the speedup with respect to the canonical serial version (**panmixia** columns of Table 3.7). But it is not fair to compute speedup against a sequential GA, since we compare different algorithms (the parallel code is that of a cGA). Hence, we compare the same algorithm (the cGA) both in sequential and in parallel (cGA$n$ on 1 versus $n$ processors). This speedup is known as orthodox speedup. The speedup, the efficiency, and the serial fraction using the orthodox definition are shown in **orthodox** columns of Table 3.7. The orthodox values are slightly better than those on the panmictic ones. But the trend in both cases is similar (in some other problems the trend could even be different); the speedup is quite good but it is always sublinear and it slightly moves away from the linear speedup as the number of CPUs increases. That is, when we increment the number of CPUs we have a moderate loss of efficiency. The serial fraction is quite stable although we can notice a slight reduction of this value as the number of CPUs increases, indicating that maybe we can gain some efficiency adjusting the granularity of the parallel task.

**Table 3.7** Speedup and Efficiency.

| **Alg.** | panmixia | | | orthodox | | |
|---|---|---|---|---|---|---|
| | speedup | efficiency | serial fract. | speedup | efficiency | serial fract. |
| **cGA2** | 1.83 | 0.915 | 0.093 | 1.91 | 0.955 | 0.047 |
| **cGA4** | 3.30 | 0.825 | 0.070 | 3.43 | 0.857 | 0.055 |
| **cGA8** | 6.50 | 0.812 | 0.032 | 6.77 | 0.846 | 0.026 |
| **cGA16** | 11.65 | 0.728 | 0.025 | 12.01 | 0.751 | 0.022 |

### *3.4.6 Example 6: A Predefined Effort Could Hinder Clear Conclusions*

In the previous examples the stopping criterion is based on the quality of the solution. In this experiment, the termination condition is based on a predefined effort (60,000 evaluations). Therefore, we focus on the quality of found solutions. Previously, we used the fitness of the best solution, and the average fitness to measure the quality of the solutions. We now turn to use a different metric, such as the median and the average of the final population mean fitness of each independent run (**mm**). In Table 3.8 we list all these metrics for a sequential GA and a distributed GA using four processors.

**Table 3.8** Result of Example 6.

| **Alg.** | % hit | best | avg | median | mm |
|---|---|---|---|---|---|
| **Seq.** | 0% | 418 | 406.4 | 401 | 385.8 |
| **dGA4** | 0% | 410 | 402.3 | 405 | 379.1 |

Using a predefined effort as a stopping criterion is not always a good idea in parallel metaheuristics if one wishes to measure speedup: in this case, for example, algorithms could not find an optimal solution in any execution. If we analyze the best solution found or the two averages (average of the best fitness, **avg** column, and average of the mean fitness, **mm** column), we can conclude that the sequential version is more accurate than the parallel GA. But the median value of dGA is larger than this of the serial value, indicating that the sequential algorithm obtained several very good solutions but the rest of them had a moderate quality, while the parallel GA had a more stable behavior. With this stopping criterion, it is hard to obtain a clear conclusion if the algorithm is not stable; in fact, normal distribution of the resulting fitness is hardly found in many practical applications, a disadvantage for simplistic statistical claims.

Also, we can notice that the **avg** data are always better than the **mm** values. This is common sense, since the final best fitness is always larger than the final mean fitness (or equal when all the individuals converge to the same single solution).

## 3.5 Conclusions

This chapter considered the issue of reporting experimental research data for parallel genetic algorithms. Since this is a difficult task, the main issues of an experimental design are highlighted. We do not enter the complex and deep field of pure statistics in this chapter (not our focus), but just some important ideas to guide researchers in their work. Readers interested in this could benefit form using ideas and procedures from `http://www.keel.es` [105].

As it could be expected, we have focused on parallel performance metrics that allow to compare parallel approaches against other techniques of the literature. Speedup and total wall-clock time are by far the most used measures to analyze pGAs. We stated the utilization of average times to represent the execution of a pGA, although other measure might be used (in fact time is not following a normal distribution and the mean could be a better value for characterizing a pGA). This is a matter for future research. Besides, we have shown the importance of the statistical analysis to support our conclusions, also in the parallel metaheuristic field.

Finally, we have performed several experimental tests to actually illustrate the influence and utilizations of the many metrics described in the chapter.

# Part II

# Characterization of Parallel Genetic Algorithms

# 4

# Theoretical Models of Selection Pressure for Distributed GAs

*My goal is simple. It is complete understanding of the Universe, why it is as it is, and why it exists at all.*

*Stephen Hawking (1942 - ) - English scientist*

The increasing availability of clusters of machines has allowed the fast development of pGAs [6]. Most popular parallel GAs split the whole population into separate subpopulations that are dealt with independently (islands). A sparse exchange of information among the component subalgorithms leads to a whole new class of algorithms that do not only perform faster (more steps by unit time), but that often lead to superior numerical performance [30, 106, 107].

Many interesting parallel issues can be defined and studied in pGAs (speedup, efficiency, etc.), but in this chapter we are interested in the underlying distributed algorithm model using multiple populations, that is really the responsible of the features of the search. From the start, we want to reinforce the difference between the model (dGA) and its implementation (pGA). In this chapter we concentrate on the dynamics of the distributed GA, in particular in developing a mathematical description for the takeover time, i.e., the time for the best solution to completely fill up all the subpopulations of the dGA. We first will propose and analyze several models for the induced growth curves as an interesting contribution by themselves, and then address the calculation of the takeover time. In this work, stochastic universal sampling selection is considered. Also, since we only focus on selection (i.e., no variation operators), we expect an easy extension of the results to many other evolutionary algorithms (EAs) different from dGAs.

In order to design a distributed GA we must take several decisions. Among them, a chief decision is to determine the migration policy: topology (logical links between the islands), migration rate (number of individuals that undergo migration in every exchange), migration frequency (number of steps in every subpopulation between two successive exchanges), and the selection/replacement of the migrants. The values of these parameters have an important influence on the algorithm behavior. In general, decisions on these choices are made by experimental studies. Therefore, it would be interesting if we could provide an analytical basis for such decisions.

G. Luque and E. Alba: Parallel Genetic Algorithms, SCI 367, pp. 55–71.
springerlink.com 

Several works have studied the takeover time and growth curves for other classes of structured EAs in the past [108, 109, 110, 111, 112, 113, 114]. In general, these works are oriented to study cellular EAs (with the important exception of the Sprave's one [114] that has a much broader coverage). As a consequence, it really exists a gap in the studies about dEAs; filling this gap will benefit to many researchers. In [115] we made a seed contribution by analyzing the effects of the migration frequency and migration rate in the growth curves and takeover times. We now unify and extend this by studying the influence of the migration topology and proposing new more accurate models.

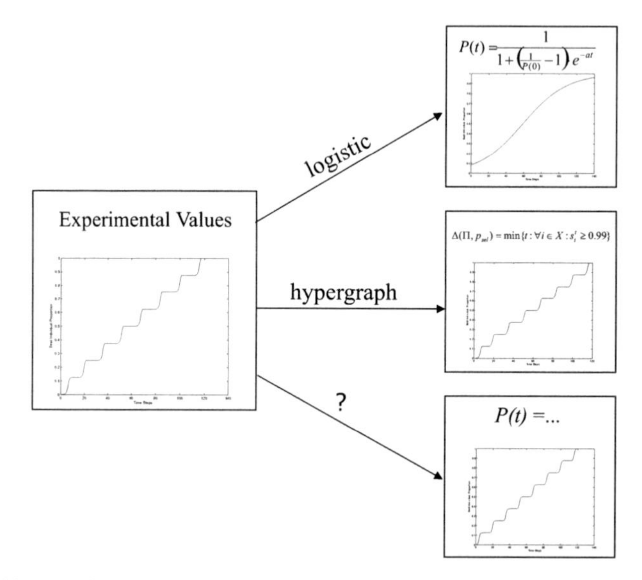


**Fig. 4.1** Sketch of our approach: study of logistic, hypergraph, and possibly other more accurate and simple models.

In the present chapter we focus on the influence of migration rate, migration frequency, and topology in the takeover time and in the growth curves. To achieve this goal we preset the policies of selection/replacement of the migrants. The emigrants are selected by binary tournament while the immigrants are included in the target population only if they are fitter than the worst-existing solution. In our analysis we will use the stochastic universal sampling selection method, plus an elitist replacement; concretely, we use a $(\mu + \mu)$-dGA. We will propose several new mathematical models for the dynamics of selection that allow to improve on the accuracy (low error) of the existing models and consequently compute more accurate takeover times (see Fig. 4.1).

This chapter is organized as follows. Section 4.1 is an introduction containing some preliminary background about previous works. In Section 4.2 we present the models that we use in this chapter. Section 4.3 studies the effects of several parameters of the migration policy in the resulting growth curves; concretely we analyze the migration frequency, the migration rate, and the migration topology. In Section 4.4, we analyze the predicted takeover times provided by the models. In the last section we summarize the conclusions and give some hints on the future work.

## 4.1 Existing Theoretical Models

A common analytical approach to study the selection pressure of an EA is to characterize its takeover time [116], i.e., the number of generations it takes for the best individual in the initial population to fill the entire population under selection only. The growth curves are another important issue to analyze the dynamics of the dEAs. These growth curves are functions that map the generation step of the algorithm to the proportion of the best individual in the whole population. In this section we briefly describe the main models found in the literature defining the behavior of structured population EAs, since our target algorithms (dEAs) are a subclass of these.

### *4.1.1 The Logistic Model*

Let us begin by discussing the work of Sarma and De Jong [109] for cellular EAs. In that work, they performed a detailed empirical analysis of the effects of the neighborhood size and shape for several local selection algorithms. They proposed a simple quantitative model for cellular EAs based in the logistic family of curves already known to work also for panmictic EAs [116]. In summary, the proposed equation is:

$$P(t) = \frac{1}{1 + \left(\frac{1}{P(0)} - 1\right) e^{-at}} \tag{4.1}$$

where $a$ is a growth coefficient and $P(t)$ is the proportion of the best individual in the population at time step $t$. This model threw accurate results for synchronous updates of square shaped cellular EAs. Recently, for the asynchronous case, improved models has been proposed in [113] not following a logistic growth. Anyway, using a logistic curve represents an interesting precedent that however should be first validated for dEAs. In brief, we will do so in this chapter.

### *4.1.2 The Hypergraph Model*

On the other hand, Sprave [114] has proposed a unified description for any non-panmictic population structured EA, that could even end in an accurate model for panmictic populations (since they can be considered as fully connected structured populations). Sprave modelled the population structure by means of *hypergraphs*. A hypergraph is an extension of a canonical graph. The basic idea of a hypergraph is the generalization of edges from pairs of vertices to arbitrary subsets of vertices (see two examples in Fig. 4.2).

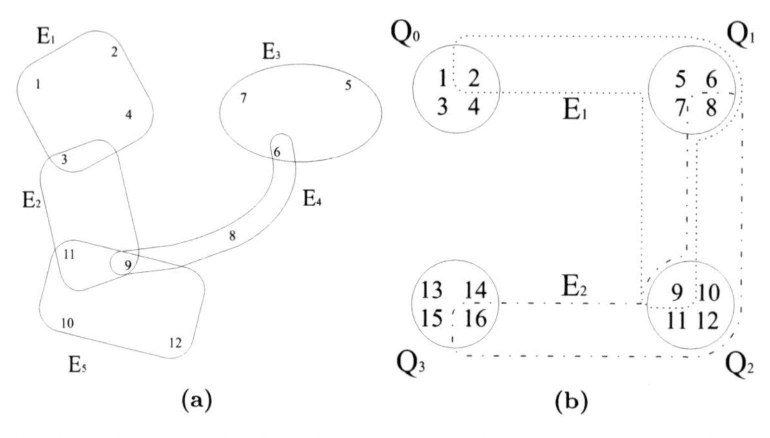


**Fig. 4.2** Two hypergraphs: (a) a basic hypergraph with vertices $X = \{1, \ldots, 12\}$ and several edges $\{E_1, \ldots, E_5\}$ and (b) another hypergraph example that represents a dEA with four subpopulations in a ring ($E_0$ and $E_3$ have been omitted for clarity).

In this work, the author developed a method to estimate growth curves and takeover times. This method was based on the calculation of the diameter of the actual population structure and on the probability distribution induced by the selection operator. In fact, Chakraborty et al. [117] previously calculated the success probabilities ($p_{select}$) for the most common selection operators, what represents an interesting complement for putting hypergraphs to work in practice.

### *4.1.3 Other Models*

Although the logistic model is relatively well-known, and hypergraphs could potentially play an important role in the field, they are not the only existing models that can inspire or influence the present study. Gorges-Schleuter [110] also accomplished a theoretical study about takeover times for a cellular ES algorithm. In her analysis, she studied the propagation of information over time through the entire population. She finally obtained a linear model for a ring population structure and a quadratic model for a torus population structure.

In a different work, Rudolph [111] carried out a theoretical study on the takeover time in populations with array and ring topologies. He derived lower bounds for arbitrary connected neighborhood structures, lower and upper bounds for array-like structures, and an exact closed form expression for a ring neighborhood structure.

Later, Cantú-Paz [9] studied the takeover time in dGAs where the migration occurs in every iteration, which is the lower bound of the migration frequency value. He generalized the panmictic model presented by Goldberg and Deb (1991) by adding a policy-dependent term. That term represents the effect of the policy used to select migrants and the individuals that they replace at the receiving island.

Finally, Giacobini et al. [112, 118] studied the takeover time in cellular EAs that use asynchronous cell update policies. The authors presented quantitative models for the takeover time in asynchronous cellular EAs with different topologies.

## 4.2 Analyzed Models

In the present work, we just focus on the two described models: logistic and hypergraphs. The first one (logistic) is based on biological processes and it is well-known in the case of cellular EAs, the other type of structured EAs. The second model (hypergraphs) possesses a unique unified-like feature for all non-panmictic algorithms. We do not use the results of the other mentioned works [9, 110, 111, 112] directly, since either they are linked to specialized algorithms (non-canonical) or have a different focus (e.g., selection policy).

Let us first address the logistic case:

$$P(t) = \frac{1}{1 + a \cdot e^{-b \cdot t}} \tag{4.2}$$

To strictly adhere to the original work of Sarma and De Jong for cellular EAs, the $a$ parameter should be defined as a constant value ($a = \frac{1}{P(0)} - 1$). We call this model LOG1. We propose a new variant of the logistic model called LOG2. In this case, we consider $a$ and $b$ as free variables (in the previous model LOG1 $a$ was a constant parameter).

We now turn to consider the hypergraph approach. In fact, we present two variants of hypergraphs: the one in which $p_{select}$ (Equation 4.3) accounts only for the probability of selection (HYP1), and the one where this probability (Equation 4.4) accounts both for selection and for replacement (HYP2). We introduce such distinction because in the seminal work [117] this second choice (combining selection and replacement within a probability) is said to be more exact (although not addressed there):

$$p_{select1}(i, M) = \frac{i \cdot \gamma}{i \cdot \gamma + M - i} \tag{4.3}$$

$$p_{select2}(i, M) = \frac{i}{M} + \left(1 - \frac{i}{M}\right) p_{select1}(i, M) \tag{4.4}$$

where $M$ is the population size and $i$ denotes the total number of best individuals in the population. These functions assume that the population has only two types of individuals where the fitness ratio is $\gamma = f_1/f_0$ [117]. In our experiments we preset $\gamma = 2$ ($\gamma = f_{best}/f_{avg(no\_best)} = 2$).

To end this section we introduce our new proposed models in details. In our previous work [115], we presented a more accurate extension of the original logistic model ($freq$ is the migration frequency and $N$ is the number of islands):

$$P(t) = \sum_{i=1}^{i=N} \frac{1/N}{1 + a \cdot e^{-b \cdot (t - freq \cdot (i-1))}} \tag{4.5}$$

but this model does not include information about the topology in the distributed EA, in fact, it only works appropriately with a ring topology. Therefore, we here propose a new extension of this model that accounts the migration topology:

$$P(t) = \sum_{i=1}^{i=d(T)} \frac{1/N}{1 + a \cdot e^{-b \cdot (t - freq \cdot (i-1))}} + \frac{N - d(T)/N}{1 + a \cdot e^{-b \cdot (t - freq \cdot d(T))}} \tag{4.6}$$

where $d(T)$ is the length of the longest path between any two islands (diameter of the topology). This expression is a combination of the logistic model plus our previous model (Equation 4.5. In fact, in the panmictic case [116] ($d(T) = 0, freq = 0$, and $N = 1$), this equation is the same as the logistic one; and if the topology is the ring one ($d(T) = N - 1$) then this model reduces to the expression shown in the Equation 4.5. We should notice that since it is an extension of the logistic approach, two variants could be also defined as we did before with LOG1 and LOG2. The first (called TOP1) in which $a$ is constant ($a = \frac{1}{P(0)/N} - 1$), and the second (TOP2) where $a$ and $b$ are adjustable parameters.

## 4.3 Effects of the Migration Policy on the Actual Growth Curves

In this section we experimentally analyze the effects of the migration policy over the growth of the best individual copies in dEAs. In this aim we begin by performing an experimental set of tests for several migration topologies, frequencies, and rates. First, we describe the parameters used in these experiments, and later we analyze the results.

### 4.3.1 Parameters

We have performed several experiments to serve as a baseline data set for evaluating the mathematical models. In these experiments we use different values of migration topologies (island, star, and fully connected), frequencies (1, 2, 4, 8, 16, 32, and 64 generations) and rates (1, 2, 4, 8, 16, 32 and 64 individuals). In general, researchers use some of these values to configure theirs dEAs. Notice that a low frequency value (e.g., 1) means high coupling, while a high value (e.g., 64) means loose coupling (large gap). First, we analyze every parameter separately (the rest of parameters are kept constant) and latter, we study all them together. In the experiments, we use a $(\mu + \mu)$-dEA with 8 islands (512 individuals per island), and stochastic universal sampling selection.

For all tests, we use randomly generated populations with individual fitness between 0 and 1023. Then we introduce a single best individual (fitness = 1024) in a uniformly randomly selected island. For the actual curves we have performed 100 independent runs.

$$MSE(model) = \frac{1}{k}\sum_{i=1}^{k}(model_i - experimental_i)^2 \tag{4.7}$$

In order to compare the accuracy of the models we proceeded to calculate the mean square error (Equation 4.7) between the actual values and the theoretically predicted ones (where $k$ is the number of points of the predicted curve). The MSE gives the error for an experiment. But we also define a metric that summarizes the error for all the experiments, thus allowing to perform a quantitative comparison between the different models easily. We initially studied several statistical values (mean, median, standard deviation, etc.) but finally we decided [115] to use the $\|\cdot\|_1$ (1-norm, Equation 4.8 where $E$ is the number of experiments) that represents the area below the MSE curve.

$$\|model\|_1 = \sum_{i=1}^{E}|MSE(model)| \tag{4.8}$$

### 4.3.2 Migration Topology

Let us begin by analyzing the curves that have been obtained in the experiments. Fig. 4.3 contains the lines of the actual takeover time for different migration topologies: ring, star, and fully connected ones (we preset moderate migration rate (8 individuals) and frequency (16 generations)). This figure shows that the migration topology mainly influences *the number of plateaus* (continuous to discrete increment in proportion of the best individual) . First, all the topologies have a small step (the islands having the best initial solution converge quickly) and when migration occurs, the best individual proportion increases again attending to the number of neighbors of the island.

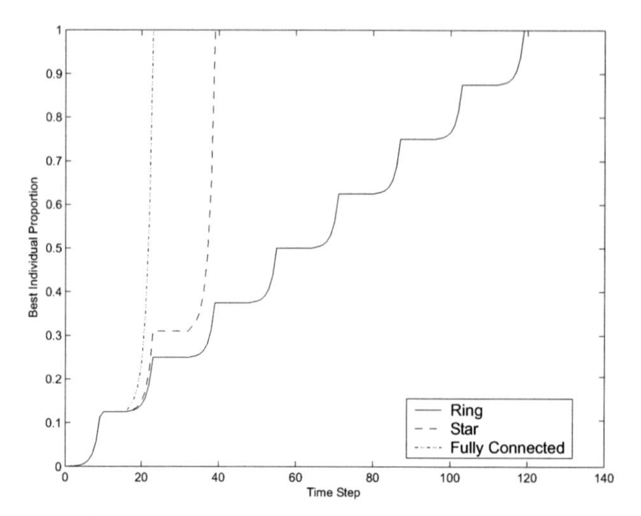


**Fig. 4.3** Actual growth curves for several migration topologies (100 independent runs).

Once we have understood the basic regularities behind, our goal is to find a mathematical model that allows an accurate fitting to all these curves.

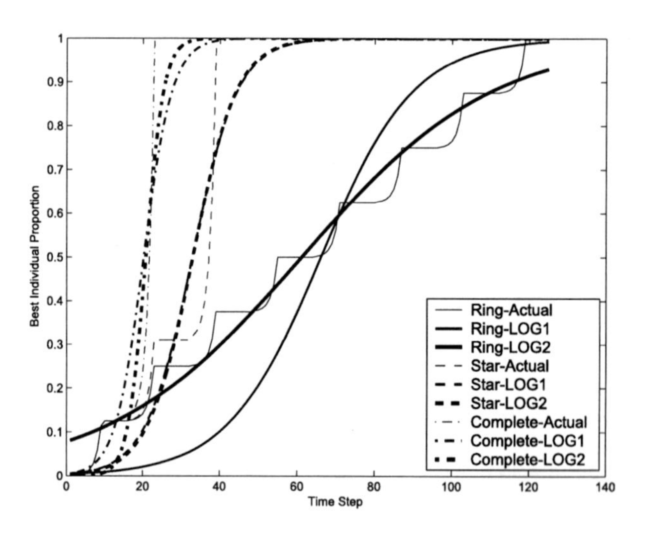


**Fig. 4.4** Comparison between actual/predicted values with LOG1 and LOG2.

We begin this task by trying to use the mentioned logistic and hypergraph models. Let us first address the logistic case. We plot its accuracy in Fig. 4.4. We can quickly arrive to the conclusion that for fully connected topology (the most similar to panmictic-like scenario), the error is small, what means good news for a logistic fitting. However, for other topologies (specially for ring topology), they turn to be very inaccurate. LOG2 allows a fitting with a smaller error than LOG1, but it still seems harmful, since the actual plateaus shown in Fig. 4.3 are ignored both in LOG1 and LOG2.

Our first clear conclusion is that the basic logistic model, even when enhanced, cannot be used for dEAs, as the existing literature also claims for other non canonical cellular EAs [110, 111, 112, 113].

Therefore, we now turn to consider the hypergraph approach. When the hypergraph model is put to work, we can notice a clear improvement over the logistic models, obtaining a very accurate curve fitting prediction with respect to the actual growth curves (Fig. 4.5). As expected, HYP1 generates a slightly worse fit than HYP2, because HYP1 is not accounting for the replacement effects.

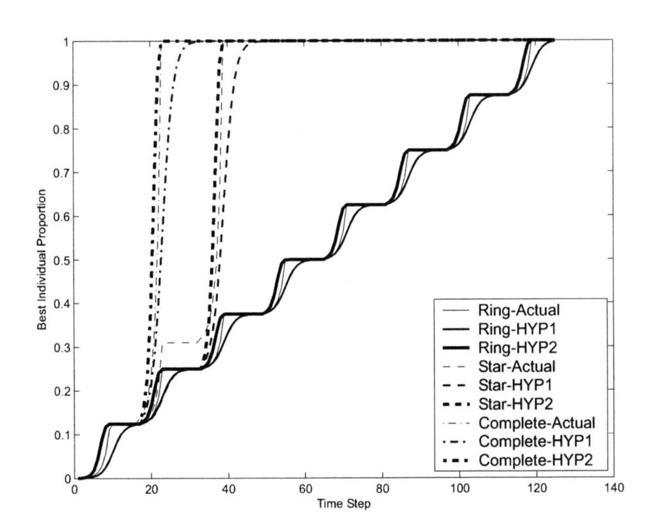


**Fig. 4.5** Comparison between actual/predicted values with HYP1 and HYP2.

To end this section we test our proposed models: TOPx. In Fig. 4.6 we plot the behavior of such models. Both models, TOP1 and TOP2, are very accurate and they have a similar behavior. They two are equally or even more precise than any other existing model, in particular with respect to the basic logistic and hypergraph variants.

### *4.3.3 Migration Frequency*

In the previous section we set the migration frequency to 16 generations. Now we perform several experiments with different values of the migration

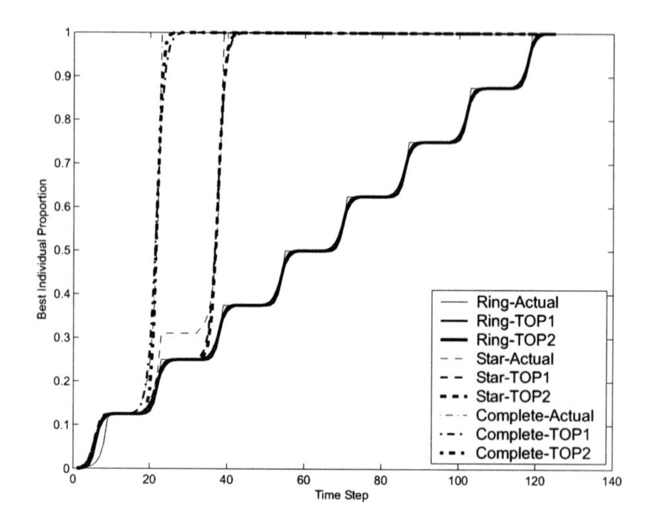


**Fig. 4.6** Comparison between actual/predicted values with TOP1 and TOP2.

frequency: 1, 2, 4, 8, 16, 32, and 64 generations (we preset moderate migration rate (8 individuals) and ring topology). Fig. 4.7 contains the lines of the actual takeover time for different migration frequencies. This figure shows that, for low frequency values, the dEA resembles the panmictic case [116]. This is common sense since there exists high interaction among the subalgorithms. However, for higher frequency values (slightly coupled search), the observed behavior is different: the subpopulations in the islands having the best solution converge quickly, and then the global convergence of the algorithm stops progressing (flat lines) until a migration of the best individual takes place. The observed effect is that of a *stairs*-like curve. The *time span* of each step in such a curve is governed by the migration frequency. The higher the migration frequency value, the largest the span of the plateau.

### 4.3.4 Migration Rate

After analyzing the topology and migration frequency, we proceed in this subsection to study the influence of the migration rate over the takeover regime. As done before, we first compute the proportion of the best individual in a dEA when utilizing the following values of the migration rate: 1, 2, 4, 8, 16, 32, and 64. The rest of the parameters are similar to the presented in Subsection 4.3.1.

In Fig. 4.8 we plot the way in which the migration rate influences the selection pressure. From this figure we can infer that the value of the migration rate determines *the slope* of the curve. The reason is that, when the migration rate value is high, the probability of migrating the best individual increases, and then the target island converges faster than if the migration rate were smaller.

A complete analysis of migration rate and frequency can be found in [115].

### *4.3.5 Analysis of the Results*

In the previous subsections, we analyzed the effects of the migration frequency, rate, and topology over the growth curves of a dEA. However, these studies only consider one parameter while the rest of them are fixed. Now, in this subsection we will try to answer a common sense question: are the results

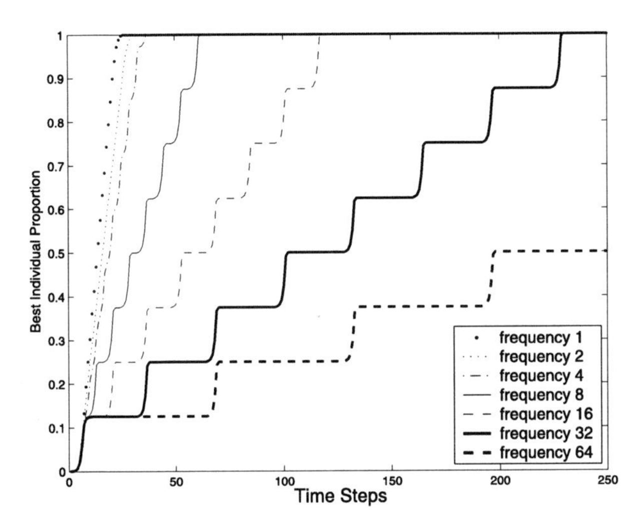


**Fig. 4.7** Actual growth curves for several migration frequencies.

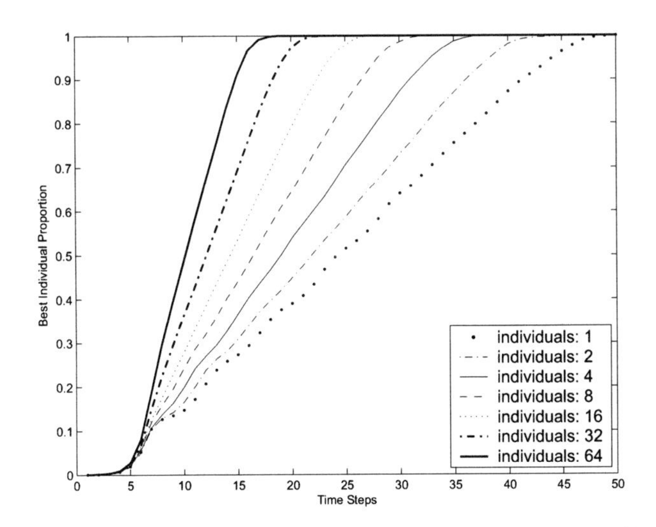


**Fig. 4.8** Actual growth curves of a dEA for several migration rates (100 independent runs).

somehow biased by the utilization of the selected parameters? Therefore, To extend the previous studies we now analyze the effects of all the parameters together.

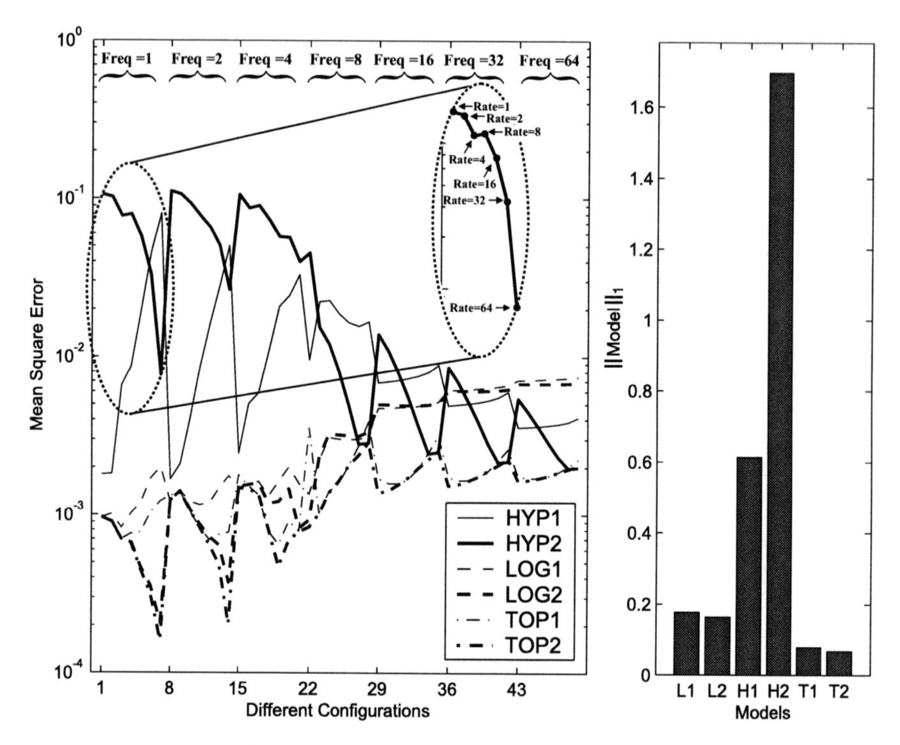


**Fig. 4.9** Error (MSE) and $\|\cdot\|_1$ between the actual and predicted growth curves for star topology and all the values of migration rate and frequency. In the right graph, Lx means LOGx, Hx means HYPx, and Tx means TOPx.

Let us now proceed with the fitting of these curves with all the considered mathematical models. In Fig. 4.9 (left) we show the error for a star topology and any combined values of the migration frequency and rate. To interpret the left graph you must notice that the first seven points of a line correspond to the error incurred by the associated predictive model for the seven different values of the migration rate at the same frequency, and that there exist seven groups of such points, one for each migration frequency from 1 to 64 (from left to right in the horizontal axis).

We can see in Fig. 4.9 some behavioral patterns of the models with respect to the final MSE error they exhibit. First, we observe that the logistic behavior is very stable and very accurate for all configurations. Both logistic variants have similar behavior although LOG1 is slightly worse than LOG2. Second, the hypergraph model obtains low error for larger frequency values, while their inaccuracy is more evident for smaller values of the migration frequency (tight coupling). Even our proposed TOPx model also is somewhat sensitive to low frequency values, but it is quite stable and accurate for larger values of migration frequency. Both, the HYPx and the TOPx models seem to perform a cycle: reduction/enlargement (respectively) of error as the migration rate enlarges (for any given frequency). The $\|\cdot\|_1$ summarizes quantitatively the MSE results in a single value per model (Fig. 4.9 right). Although the HYPx models are very accurate for lager frequency values, they show high errors for smaller values, thus making the $\|\cdot\|_1$ metric larger than the rest. The HYPx models are not as robust as expected. Also, LOGx obtain very accurate results and are only worse than the TOPx models. Clearly, our TOP2 obtains the lowest overall error.

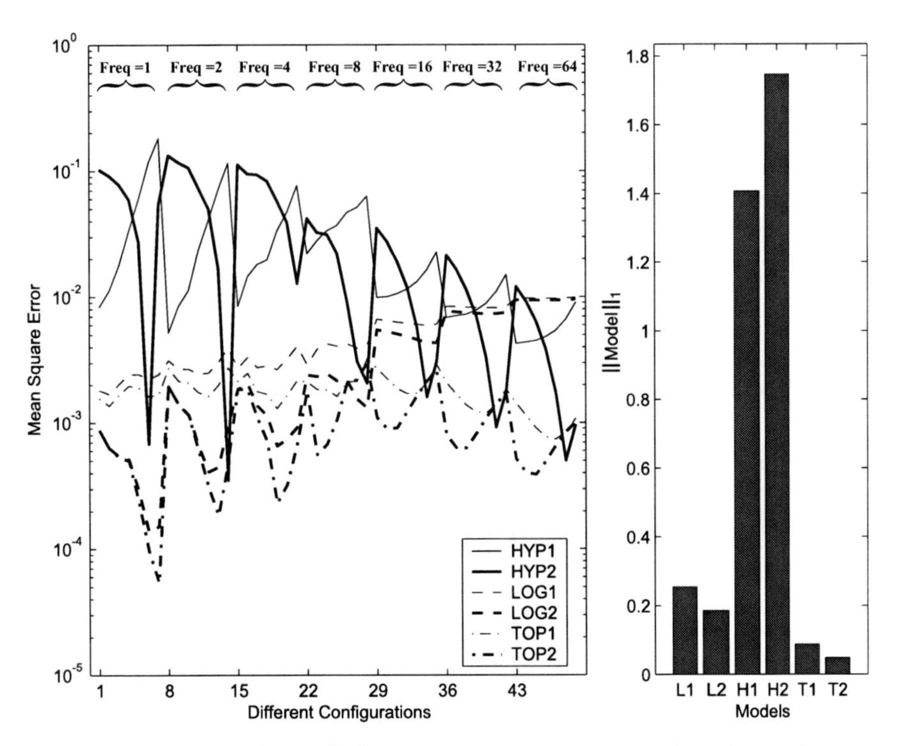


**Fig. 4.10** Error (MSE) and $\|\cdot\|_1$ between the actual and predicted growth curve for fully connected topology and all the values of migration rate and frequency.

In figures 4.10 and 4.11 we show the error for fully connected and ring topologies, respectively. First, we notice that the behavior of the models is very similar to that showed previously for the star topology (Fig. 4.9), i.e., HYPx models are the most inaccurate, while LOGx are quite precise and they only are outperformed by TOPx models. On the other hand, the behavior of the models showed in Fig. 4.11 (ring) is quite different from previous ones.

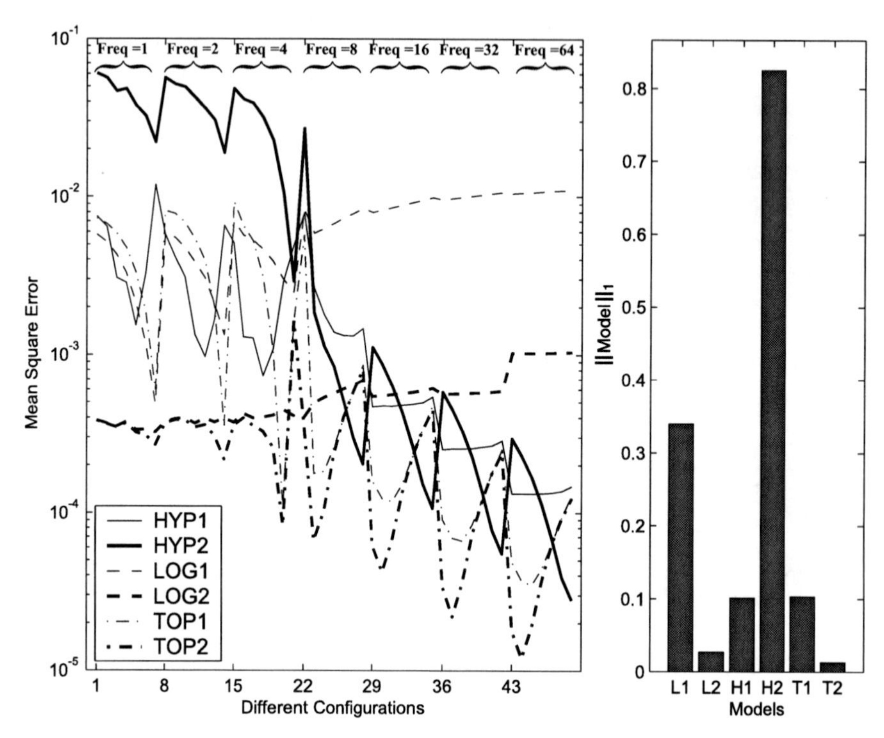


**Fig. 4.11** Error (MSE) and $\|\cdot\|_1$ between the actual and predicted growth curve for ring topology and all the values of migration rate and frequency.

We can notice that in the case of low values of the migration frequency, most of the models obtain a large error (with the exceptions of LOG2 and SUM2 models). In general, the second variant of logistic and our proposed models are always better than the first one. The LOGx models show a stable behavior for any frequency, but while LOG1 is always very inaccurate, the LOG2 performs well, and it is only worse than TOP2 model. The TOPx and HYPx models reduce their errors as the migration frequency enlarges, although the TOPx models are always more accurate than the HYPx ones. In both figures, TOP2 is again the overall most accurate model.

## 4.4 Takeover Time Analysis

In the previous sections we have studied the effects of the migration gaps and migration rate over the takeover growth curves. Now, we analyze the effect of these parameters over the takeover time itself. Fig. 4.12 contains the value of the actual takeover time for different migration frequencies, rates, and topologies. We can notice that the takeover value increases for higher frequencies and smaller rates, as expected from a loosely coupled set of subalgorithms. However, the rate effect over the takeover time is smoother (less influential) than the frequency one. Also, the takeover value increases depending on the diameter of the topology, e.g., the ring topology has the largest diameter ($d(T) = N - 1$) and then its convergence is slower.

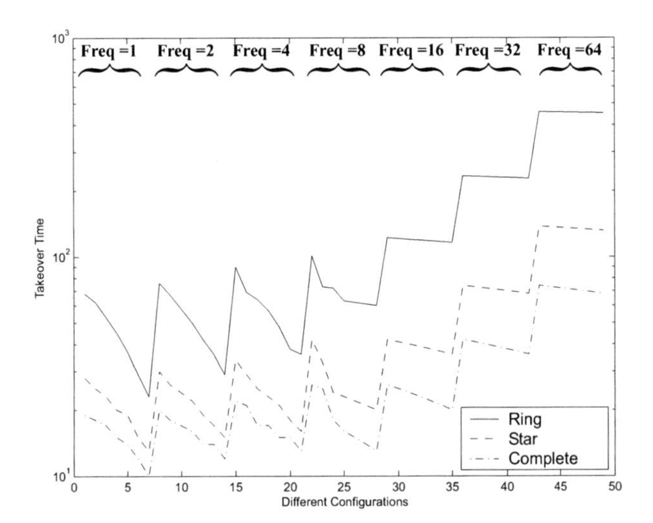


**Fig. 4.12** Actual takeover time values for all configurations.

Once we have observed the effect of the migration rate and frequency over the actual takeover time, we analyze the predicted values provided by the models. Fig. 4.13 shows the mean error of the predicted takeover time with all the models. To find these takeover time values, we numerically iterate the formula until it reaches 1. Specially interesting is the case of LOGx models, that obtain a very accurate fitting of the growth curves but are quite inaccurate to predict the takeover time. On the contrary, the HYPx models obtains the best predictions of the takeover values, but they are not useful to calculate the growth curves as shown before. Our proposed models, TOPx, obtain slightly worse prediction than HYPx for the takeover time, but in general these differences are not significant. Also, we can notice that the predictions of the models are quite sensitive to the topology; the longest the migration topology diameter, the largest the error of models.

Finally, we conclude this section by showing a closed equation for the takeover time calculation for the new models presented in this chapter (TOPx models) and an evaluation of its accuracy. The formula of takeover time (Equation 4.9) is derived from the growth curve equation (Equation 4.6) of these models, finding the convergence time of the last island that receives the best individual, i.e., finding the $t$ value for which the second term of the equation is evaluated to $\frac{N-d(T)}{N} - \varepsilon$:

$$t^* = freq \cdot d(T) - \frac{1}{b} \cdot Ln\left(\frac{1}{a} \cdot \frac{\varepsilon}{N - d(T) - \varepsilon \cdot N}\right) \tag{4.9}$$

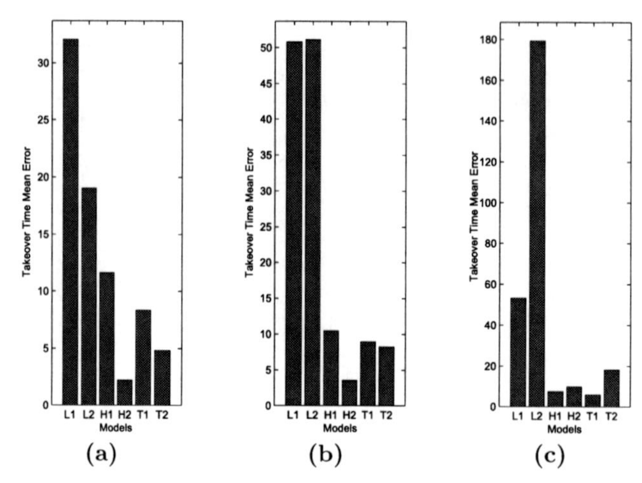


**Fig. 4.13** Error between actual and predicted values of takeover time for (a) fully connected, (b) star, and (c) ring topologies.

where $t^*$ is the takeover time value, $freq$ is the migration frequency, $N$ is the number of islands, $d(T)$ is the topology diameter, and $\varepsilon$ is the the expected level of accuracy (a small value near zero).

In Fig. 4.14 we show the error obtained when we use the above formula to estimate the takeover time given the values of migration frequency, migration rate, and migration topology. The predicted values are quite accurate and the mean error is very low. We also can observe that the errors obtained with the Equation 4.9 are significant smaller than the ones obtained when the TOPx growth models are iterated.

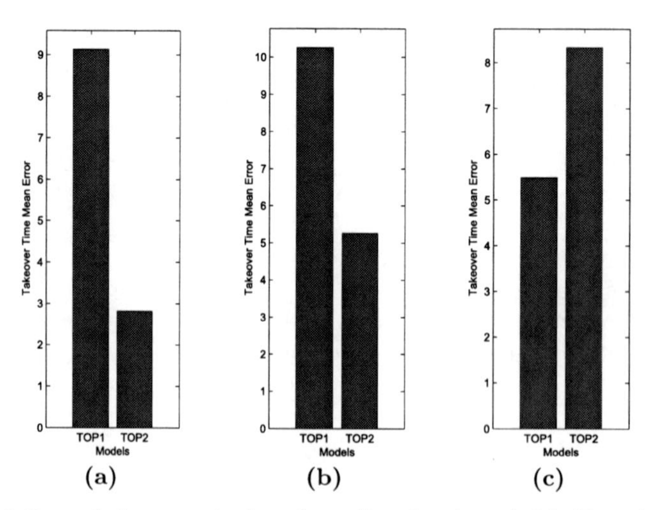


**Fig. 4.14** Error between actual and predicted values (with Equation 4.9) of takeover time for (a) fully connected, (b) star, and (c) ring topologies.

## 4.5 Conclusions

In this chapter we have performed an analysis of the growth curves and takeover regime of distributed genetic algorithms. We compared the well-known logistic model, a hypergraph model, and a newly proposed model. A second variant of each model has been also developed for the shake of accuracy.

In this chapter we have shown how the models appropriately captured the effects of the most important parameters of the migration policy: migration frequency, rate, and topology. We have observed that the migration topology affects to the number of plateaus, the migration frequency establishes the size of these plateaus, and the migration rate indicates the slope of the growth curves. These mathematical models have also allowed to study the importance of the different migration parameters in the takeover time. The main conclusion is that the influence of the migration rate is negligible when a medium/large migration frequency value is used.

Although every model has its own advantages, either simplicity (LOGx), extensibility (HYPx), or accuracy (TOPx), TOP2 is the model that better fits the actual observed behavior of the algorithms, and its predicted takeover time does not show significant differences with respect to the rest.

As a future work we plan to check the results presented in this chapter on additional selection methods and structured models of GAs (e.g., cellular GAs). Other interesting open research line (in which we are already working) is to get an statistical model of the dGA and then use it to complete the proposed models of this chapter with that information in order to obtain a more accurate mathematical description. The preliminary results are very promising [71]. Another important future work is to apply these mathematical model to help the method during the search. For example, these models can be used to build an self-adaptive method where some migration parameters could be adjusted automatically (online during the search) to adapt the behavior of the technique to the search space of the problem based on a theoretical prediction on what will happen in the next few iterations.

# Part III

# Applications of Parallel Genetic Algorithms

# 5

# Natural Language Tagging with Parallel Genetic Algorithms

*Evolution is cleverer than you are.*

*Francis Crick (1916 - 2004) - English scientist*

This is the first in a series of four chapters focused to illustrating how pGAs can be applied in a wide set of difficult tasks with success. The target domains will be Natural Language, hardware design, scheduling, and Bioinformatics, all of them hot topics in present research. Part-of speech (POS) tagging or simply "tagging" is a basic task in natural language processing (NLP). Tagging aims to determine what is the most likely lexical tag for a particular occurrence of a word in a sentence. There are many NLP tasks which can be improved by applying *disambiguation* to the text [119]. The ambiguity of syntactic analysis or partial parsing –a kind of analysis limited to particular types of phrases– is highly simplified in the absence of lexical ambiguity. Many partial parses work on the output of a tagger [120] by testing the appearance of regular expressions of tags, which define the searched patterns. For instance, the word *can* can be a noun, an auxiliary verb or a transitive verb. The category assigned to the word will determine the structure of the sentence in which it appears and thus its meaning.

Let us consider some examples, taken from newspaper headlines, in which the meaning is dramatically altered when different tags are assigned to some words:

Soviet virgin *lands* short of goal again

British *left waffles* on Falkland islands

Teacher *strikes idle* kids

We can observe that the meaning of the sentences is very different depending on the words marked in italic being noun, verb or adjective.

Other important applications of tagging are information retrieval [121] and question answering. It refers to mechanisms, such as those used by Web search engines, for easily access information items. For example, before identifying the documents relevant for the requested information, an information retrieval system needs to represent those documents according to some

G. Luque and E. Alba: Parallel Genetic Algorithms, SCI 367, pp. 75–89.
springerlink.com 

criterion, such as the set of terms which characterize them. Tagging and partial parsing can be very useful here to perform such an organization and obtain the representative terms, usually composed of more than one word, such as nominal phrases.

Tagging is also useful in question answering, which aims to answer a user query, usually with a noun phrase, such as a date, location, etc. Tagging the query helps to identify the entities the user is looking for.

Moreover, tagging is a difficult problem by itself, since many words belong to more than one lexical class. To give an idea, according to [122], over 40% of the words appearing in the hand-tagged Brown corpus [123] are ambiguous.

Because of the importance and difficulty of this task, a lot of work has been carried out to produce automatic taggers. Automatic taggers, usually based on Hidden Markov Models, rely on statistical information to establish the probabilities of each scenario. The statistical data are extracted from previously hand-tagged texts, called *corpus*. These stochastic taggers [119, 124] neither require knowledge of the rules of the language nor try to deduce them, and thus they can be applied to texts in any language, provided they can be previously trained on a corpus for that language.

The context in which the word appears helps to decide which is its more appropriate tag, and this idea is the basis for most taggers. For instance, consider the sentence in Fig. 5.1, extracted from the Brown corpus. The word *questioning* can be disambiguated as a common name if the preceding tag is disambiguated as an adjective. But it might happen that the preceding word to be ambiguous, so there may be many dependencies which must be resolved simultaneously.

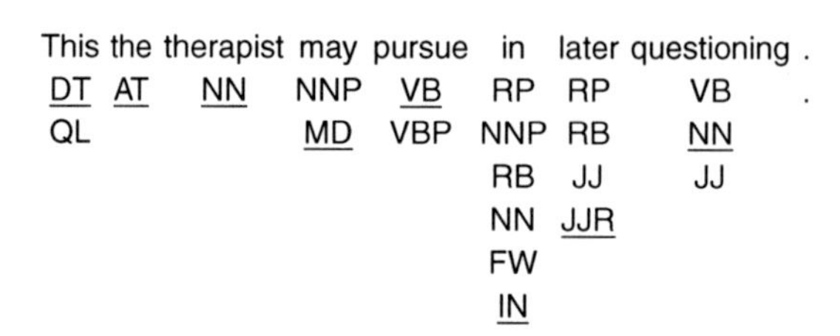


**Fig. 5.1** Tags for the words in a sentence extracted from the Brown corpus. Underlined tags are the correct ones, according to the Brown corpus. Tags correspond to the tag set defined in the Brown corpus: DT stands for determiner/pronoun, AT for article, NN for common noun, MD for modal auxiliary, VB for uninflected verb, etc.

The statistical model considered in this work amounts to maximize a global measure of the probability of the set of contexts (a tag and its neighboring tags) corresponding to a given tagging of the sentence. Then, we need a method to perform the search of the tagging which optimizes this measure of probability.

The aim of this chapter is to design a parallel GA to tackle this problem. We check and compare different (parallel and serial) metaheuristic algorithms to perform such a search, such as a genetic algorithm (GA), a non-traditional genetic algorithm: CHC method, and a simulated annealing (SA). One of the advantages of using an evolutionary algorithm (EA) as the search algorithm for tagging is that these algorithms can be applied to any statistical model, even if they do not rely on the Markov assumption (i.e., the tag of a word only depends on the previous words), as it is required in classic search algorithms for tagging, such as the widely used Viterbi [125]. Genetic algorithms have been previously applied to the problem [126, 127], obtaining accuracies as good of those of typical algorithms used for stochastic tagging. CHC is a non-traditional genetic algorithm, which presents some particular features: CHC guarantees the survival of the best solutions found, does not allow the mating of similar solutions, and uses specialized operations.

One of the aims of this work is to investigate if the particular mechanisms of CHC for diversity can improve the selection of different sets of tags. From previous work [128], it has been observed that words incorrectly tagged are usually those which require one of their more rare tags, or which appear in an infrequent context. We plan to use a quality function based on the probability of the contexts of a sequence of tags assigned to a sentence. A priori, it should be difficult for a GA to find appropriate tags within high probability contexts. CHC allows simultaneously changing several tags of the sequence, which can hopefully lead to explore combinations of tags very different from those of the ancestors and then to better results. Thus, it is interesting to study what is more advantageous; the smooth exploration of the GA or the more disruptive one of CHC. We have also compared the results of the GAs with those obtained from SA, in order to ascertain the suitability of the evolutionary approach compared with other optimization methods.

For most tagging applications, the whole process of search is time consuming, what made us to include a parallel version of the algorithms. We also compare the results of our approaches with the ones of Viterbi, a classical method for solving this problem, in order to test the accuracy of all our methods in a wider spectrum of techniques.

The rest of the chapter proceeds as follows: Section 2 describes the kind of statistical models to which our algorithms can be applied. Sections 3, 4 and 5 describe briefly the GA, CHC, and SA algorithms, and Section 6 discusses the parallel version of these algorithms. Section 7 presents the details on how the algorithms are applied to tagging. Section 8 describes and discusses the computational results, and Section 9 draws the main conclusions of this work.

## 5.1 Statistical Tagging

Statistical tagging, probably the most extended approach nowadays, is based on statistical models defined on a number of parameters, which take their values from probabilities extracted on tagged texts. The goal of these models is to assign to each word of a sentence the most likely lexical tag according to the *context* of the word, i.e., according to the tags of other words surrounding the considered one. Therefore, we can collect statistics on the number of occurrences of the contexts resulting of assigning their different valid part-of-speech to the considered word, and then choose the more likely one. However, the surrounding words may also be ambiguous, and thus, we need some kind of statistical model to select the "best" tagging for the whole sequence according to the model. More formally, the part-of-speech tagging problem can be stated as

$$t_{1,n} = \arg\max_{t_{1,n}} P(t_{1,n}|w_{1,n}) \tag{5.1}$$

where $\arg\max_x f(x)$ is the value of $x$ which maximizes $f(x)$, and $t_{1,n}$ is the tag sequence of the words $w_{1,n}$ which compose the sentence being tagged.

If we assume that the tag of a word only depends on the previous tag, and that this dependency does not change throughout the time, we can adopt a Markov Model for tagging. Let $w_i$ the word at position $i$ in the text, $t_i$ the tag of $w_i$, $w_{i,j}$ the words appearing from position $i$ to $j$, and $t_{i,j}$ the tags for the words $w_{i,j}$. Then, the model states that

$$P(t_{i+1}|t_{1,i}) = P(t_{i+1}|t_i) \tag{5.2}$$

If we also assume that the probability of a word appearing at a particular position only depends on the part-of-speech tag assigned to that position, the optimal sequence of tags for a sentence can be estimated as:

$$\begin{aligned} t_{1,n} &= \arg\max_{t_{1,n}} P(t_{1,n}|w_{1,n}) = \\ &= \prod_{i=1}^{n} P(w_i|t_i)P(t_i|t_{i-1}) \end{aligned} \tag{5.3}$$

Accordingly, the parameters of the Markov model tagger can be computed from a training corpus. It can be done by recording the different contexts of each tag in a table called *training table*. This table can be computed by going through the training text and recording the different contexts and the number of occurrences of each of them for every tag in the tagset.

The Markov model for tagging described above is known as a *bigram* tagger because it makes predictions based on the preceding tag, i.e. the basic unit considered is composed of two tags: the preceding tag and the current one. This model can be extended in such a way that predictions depend on more than one preceding tag. For example, a *trigram* model tagger makes its predictions depending on the two preceding tags.

Once the statistical model has been defined, most taggers use the Viterbi algorithm [125] (a dynamic programming algorithm) to find the tag sequences which maximize the probability according to the selected Markov model.

We here offer an alternative approach to tagging which can be used instead of the Viterbi algorithm. This approach relies on using evolutionary algorithms, which include a number of search templates based on the production of offsprings and the survival of the fittest. These heuristic techniques provide us a general method that can be applied to any statistical model. For example, they can be applied to perform tagging according to the Markov model described above or not. In this days, they can also be applied to other models for which there is no known efficient algorithm. For instance, they can be applied to a model which has been proven to improve the results over a Markov one [129], in which the context of a word is composed of both, the tag of the preceding words and also the tag of the following words. The potential problem in using evolutionary algorithms (EAs) is that many of them do not guarantee to reach the optimum solution but a reasonably good approximation, according to the resources assigned (time and memory). In addition, the probability of error can be systematically decreased in EAs by increasing the number of points explored with a fine tuning of the algorithm parameters.

According to these considerations, we are going to explore different metaheuristic techniques for tagging. The statistical model we consider amounts to maximize a global measure of the probability of the set of contexts (a tag and its neighboring tags) corresponding to a given tagging of the sentence. The contexts considered are very general and widely applicable since they are composed of a certain number of tags on the left and another on the right of the word at the considered position.

## 5.2 Automatic Tagging with Metaheuristics

We here introduce the proposed algorithms to solve the tagging problem. We have used three different methods: a canonical GA, a non-traditional GA called CHC, and a SA which is a trajectory method. We have also developed parallel version for the these algorithms. All of them will be used as automatic taggers and their performance will be compared to highlight their relative strengths in this task.

### *5.2.1 Genetic Algorithm*

Genetic Algorithms (GAs) [130] are stochastic search methods that have been successfully applied in many real applications. A GA is an iterative technique that applies stochastic operators on a pool of individuals (tentative solutions). A fitness function allocates a real value to every individual indicating its suitability to the problem. Traditionally, GAs are associated to the use of a binary representation, but nowadays you can find GAs that use other types of representations. A GA usually applies a recombination operator on two solutions, plus a mutation one that randomly modifies the individual contents to promote diversity and thus reaching new portions of the search space not implicitly present in the previous generations.

### 5.2.2 CHC Algorithm

CHC [131] is a variant of a genetic algorithm with a particular way of promoting diversity. It uses a highly disruptive crossover operator to produce new individuals maximally different from their parents. It is combined with a conservative selection strategy which introduces a kind of inherent elitism. The main features of this algorithm are:

- The mating is not restricted to the best individuals, but parents are randomly paired in a mating pool. However, recombination is only applied if the Hamming distance between the parents is above a certain threshold (*incest prevention*).
- CHC uses a *half-uniform crossover* (HUX), which exchanges half of the differing genes.
- CHC guarantees survival of the best individuals selected from the set of parents and offsprings.
- Mutation is not applied directly. Instead, CHC uses a re-start mechanism when the population remains unchanged after a given number of generations.

### 5.2.3 Simulated Annealing

Simulated Annealing (SA) [12] is a stochastic search technique that can be seen as a hill-climber with an internal mechanism to escape from local optima. In SA, the solution $s'$ is accepted as the new current solution $s$ if $\delta \geq 0$ holds, where $\delta = f(s') - f(s)$. To allow escaping from a local optimum, moves that decrease the energy function are accepted with a decreasing probability $exp(\delta/T)$ if $\delta < 0$, where $T$ is a parameter called the "temperature". The decreasing values of $T$ are controlled by a cooling schedule, which specifies the temperature values at each stage of the algorithm, what represents an important decision for its application. Here, we are using a proportional method for updating the temperature ($T_k = \alpha \cdot T_{k-1}$, where $\alpha$ indicates the decrease speed of the temperature).

### 5.2.4 Parallel Versions

A parallel EA (PEA) is an algorithm composed of multiple EAs, regardless of their population structure. Each component (usually a traditional EA) subalgorithm includes an additional phase of *communication* with a set of subalgorithms [10]. In this work, we have chosen a *distributed EA* (dEA) because of its popularity and because it can be easily implemented in clusters of machines. In distributed EAs (also known as Island Model) there exists a small number of islands performing separate EAs, and periodically exchanging individuals after a number of isolated steps (*migration frequency*). Concretely, we use a static ring topology in which the best individual is migrated, and

| **Sent.** | This | the | therapist | may | pursue | in | later | questioning |
|---|---|---|---|---|---|---|---|---|
| Ind. 1: | DT | AT | NN | NNP | VBP | IN | JJ | VB |
| Ind. 2: | DT | AT | NN | MD | VB | RB | RB | NN |
| Ind. 3: | QL | AT | NN | NNP | VB | FW | JJ | JJ |

**Fig. 5.2** Potential individuals for the sentence in Fig. 5.1.

asynchronously included in the target population only if it is better than the local worst-existing solution.

The parallel SA (PSA) is also composed of multiple asynchronous SAs. Each component SA, starts off from a different random solution and exchanges the best solution found (*cooperation* phase) with its neighboring SA in the ring.

## 5.3 Algorithm Decisions: Representation, Evaluation, and Operators

The first step in designing a metaheuristic is to define the data structure included into the individuals which compose the population. Genetic operators on them must also be defined, as well as a selection policy based on a measure of the individual quality, or "fitness".

### *5.3.1 Individuals*

Tentative solutions here are made of sequences of genes. Each gene corresponds to each word in the sentence to be tagged. Fig. 5.2 shows some example individuals for the sentence in Fig. 5.1.

| **word** | **tag index** | | | | | | **int** | **bin** |
|---|---|---|---|---|---|---|---|---|
| | 0 | 1 | 2 | 3 | 4 | 5 ··· | | |
| This | DT | QL | | | | | 0 | 000 |
| the | AT | | | | | | 0 | 000 |
| therapist | NN | | | | | | 0 | 000 |
| may | NNP | MD | | | | | 1 | 001 |
| pursue | VB | VBP | | | | | 0 | 000 |
| in | RP | NNP | RB | NN | FW | IN | 5 | 101 |
| | ... | | | | | | | |

**Fig. 5.3** Integer and binary codings of a possible selection of tags chosen for the words of a sentence extracted from the Brown corpus. The selected tags appear underlined.

Each gene represents a tag and additional information useful in the evaluation of the solution, such as counts of contexts for this tag according to the training table. Each gene's tag is represented by an index to a vector which contains the possible tags of the corresponding word. The composition of the genes depends on the chosen coding, as Fig. 5.3 shows. In the integer coding the gene is just the integer value of the index. In the binary coding the gene is the binary representation of the index. As in the texts we have used for experiments the maximum number of tags per word is 6, we have used a binary code of 3 bits.

The chromosomes forming the initial population are created by randomly selecting from a dictionary one of the valid tags for each word, with a bias to the most probable tag. Words not appearing in the dictionary are assigned the most probable for its corresponding context, according to the training text.

### *5.3.2 Fitness Evaluation*

The fitness of an individual is a measure of the total correctness probability of its sequence of tags, according to the data from the training table. It is computed as the sum of the fitness of its genes, $\sum_i f(g_i)$. The fitness of a gene is defined as

$$f(g) = \log P(T|LC, RC) \tag{5.4}$$

where $P(T|LC, RC)$ is the probability that the tag of gene $g$ is $T$, given that its context is formed by the sequence of tags $LC$ to the left and the sequence $RC$ to the right (the logarithm is taken in order to make fitness additive). This probability is estimated from the training table as

$$P(T|LC, RC) \approx \frac{occ(LC, T, RC)}{\sum_{T' \in \mathcal{T}} occ(LC, T', RC)} \tag{5.5}$$

where $occ(LC, T, RC)$ is the number of occurrences of the list of tags $LC, T, RC$ in the training table, and $\mathcal{T}$ is the set of all possible tags of $g_i$.

A particular sequence $LC, T, RC$ may not be listed in the training table, either because its probability is strictly zero (if the sequence of tags is forbidden for some reason) or, most likely, because there is insufficient statistics. In these cases we proceed by successively reducing the size of the context, alternatively ignoring the rightmost and then the leftmost tag of the remaining sequence (skipping the corresponding step whenever either $RC$ or $LC$ are empty) until one of these shorter sequences matches at least one of the training table entries or until we are left simply with $T$. In this latter case we take as fitness the logarithm of the frequency with which $T$ appears in the corpus (also contained in the training table).

### *5.3.3 Genetic Operators*

For the GA, we use a one point crossover, i.e. a crossover point is randomly selected and the first part of each parent is combined with the second part of the other parent thus producing two offsprings. Then, a mutation point is randomly selected and the tag of this point is replaced by another of the valid tags of the corresponding word. The new tag is randomly chosen according to its probability (the frequency at which it appears in the corpus).

The CHC algorithm applies HUX crossover, randomly taking from each parent half of the tags in which they differ and exchanging them.

Individuals resulting from the application of the genetic operators along with the old population are used to create the new one.

## 5.4 Experimental Design and Analysis

We have used as the set of training texts for our taggers the Brown [123] and Susanne [132] corpora, two of the most widespread in linguistics. For the Brown corpus we have used a training set of 185,056 words, and a test set of 2421 words. For the Susanne corpus we have used a training set of 3213 words and a test set of 2510 words. The CHC algorithm has been run with a crossover rate of 50%, without mutation. Whenever convergence is achieved, 90% of population is renewed. The GA applies the recombination operator with a rate of 50%, and the mutation operator with a rate of 5%. In the parallel version, the migration occurs every 10 generations. We made preliminary tests with different parameter settings for determining the best values for each algorithms. The analysis of other specific operators is deferred for a future work.

**Table 5.1** Tagging accuracy obtained with the CHC algorithm for two test texts. PS stands for Population Size.

| Context | | Integer | | | | Binary | | | |
|---|---|---|---|---|---|---|---|---|---|
| | | PS = 32 | | PS = 64 | | PS = 32 | | PS = 64 | |
| | | **Seq.** | **Par.** | **Seq.** | **Par.** | **Seq.** | **Par.** | **Seq.** | **Par.** |
| Brown | 1-0 | 91.02 | 91.74 | 91.35 | 91.55 | 94.98 | **95.32** | 94.67 | 94.62 |
| | 2-0 | 91.31 | 91.31 | 91.83 | 92.18 | **95.35** | **95.35** | 95.18 | 95.03 |
| Susanne | 1-0 | 91.43 | 91.19 | 92.32 | 92.68 | 94.35 | **95.01** | 93.61 | 94.41 |
| | 2-0 | 93.42 | 93.75 | 93.53 | 93.56 | 94.37 | **94.82** | 93.96 | 94.31 |

Tables 5.1, 5.2, and 5.3 show the results obtained with CHC, GA, and SA algorithms, using both, integer and binary codings. The two upper rows correspond to the Brown text with two different contexts (1-0 is a context which considers only the tag of the preceding word and 2-0 considers the tag of the two preceding words) and the two lower rows are the results for the Susanne text. The two texts contain 2500 words approximately. Figures represent the best result out of 30 independent runs. The globally best result for each row appears in boldface. **Integer** stands for the integer representation and **Binary** for the binary representation with a code of 3 bits. For CHC

**Table 5.2** Accuracy obtained with the GA for the two test texts. PS stands for Population Size.

| Context | | Integer | | | | Binary | | | |
|---|---|---|---|---|---|---|---|---|---|
| | | PS = 32 | | PS = 64 | | PS = 32 | | PS = 64 | |
| | | **Seq.** | **Par.** | **Seq.** | **Par.** | **Seq.** | **Par.** | **Seq.** | **Par.** |
| Brown | 1-0 | 95.83 | **96.01** | 94.34 | 94.95 | 93.15 | 93.02 | 92.97 | 93.02 |
| | 2-0 | 96.13 | **96.41** | 95.14 | 95.42 | 94.86 | 94.82 | 94.54 | 94.89 |
| Susanne | 1-0 | 96.41 | **96.74** | 95.46 | 96.25 | 95.12 | 95.32 | 94.96 | 94.54 |
| | 2-0 | **97.32** | **97.32** | 96.91 | 97.01 | 95.39 | 95.43 | 95.34 | 95.39 |

and GA, measures have been taken for two population sizes. Furthermore, sequential and parallel versions with 4 islands are analyzed. In evolutionary algorithms, the population size of each island is the global population size divided by the number of islands.

Looking at Table 5.1, the first conclusion is that the binary coding always achieves a higher accuracy with respect to the integer one. This suggests that the integer representation is not appropriate for CHC, probably because the low number of genes of the latter interferes with the CHC mechanism to avoid crossover between similar individuals. Regarding the parallel executions, we can observe that the parallel version usually provides more accurate results, particularly for a population of 32 individuals, because for such a small population the higher diversity introduced by parallelism is beneficial. In general, the accuracy obtained for Brown text is better than for Susanne text, probably because the statistics provided by the Susanne corpus are poorer[1] and the CHC mechanism for diversity is limited by these data. Anyway, the best results are always obtained using binary coding and parallel executions for any instance.

Table 5.2 shows the results obtained with the GA. In this case, the integer representation provides the best results. Again, parallel versions improve the sequential results, obtaining the best results when the population is composed of 32 individuals. We can observe that for this algorithm, the accuracy increases when the context is larger. The same trend was observed before for CHC, but not so conclusively as for the GA. Also, we can notice that, unlike the previous results for CHC, the results for Susanne text are more accurate than the ones for Brown text.

Table 5.3 presents the data obtained with the SA algorithm. The SA algorithm performs 5656 iterations using a Markov chain of length 800 and with a decreasing factor of 0.99. In the parallel version, each SA component exchanges the best solution found with its neighbor SA in the ring every 100 iterations. We can observe that SA always provides worse results than any of

[1] Due to the small size of the Susanne corpus, and thus of the training set used, and also due to the large tagset of Susanne corpus.

the evolutionary algorithms, thus proving the advantages of the evolutionary approach. Anyhow, the parallel SA is still better in accuracy than the sequential version, as we also noticed for CHC and GA. In general, the results obtained for this algorithm are very poor, indicating that SA is not able to solve this problem adequately.

**Table 5.3** Accuracy obtained with the SA algorithm for the two test texts (best result out of thirty independent runs).

| Context | | Integer | | Binary | |
|---|---|---|---|---|---|
| | | **Seq.** | **Par.** | **Seq.** | **Par.** |
| Brown | 1-0 | 91.41 | **91.83** | 91.25 | 91.58 |
| | 2-0 | 91.92 | **92.28** | 91.68 | 91.72 |
| Susanne | 1-0 | 91.03 | **91.87** | 89.79 | 90.53 |
| | 2-0 | **92.31** | **92.31** | 91.31 | 91.74 |

Table 5.4 presents the best value and the average for the configuration which provides the best results of each algorithm, i.e., parallel implementation using integer coding for GA and SA, and binary one for CHC. We do not show the standard deviation because the fluctuations in the accuracy of different runs are always within the 1% interval, claiming that all the algorithms are very robust. Also, we include the results of the Viterbi method (as said before, a typical algorithm widely used for stochastic tagging) to perform a comparison between our evolutionary approach and a classical tagging method. The results of this algorithm have been obtained with the TnT system [133], a widely used trigram[2] tagger.

First we compare our algorithms, and latter, we compare our results with the Viterbi ones. We can observe in Table 5.4 that the GA has reached the globally best results for all the test texts and contexts, though the differences are small. This proves that the exploration of the search space given by the classical crossover and mutation operators are powerful enough for this specific problem.

Now, we compare the Viterbi method against the GA (which obtains the best results of all our approaches). Viterbi obtains the best results for the Brown corpus, although the difference of accuracy with respect to GA is around 1%. However, for the Susanne corpus, with provided poorer statistics, the GA results outperform Viterbi's one. In this way, the heuristic nature of the genetic algorithm is illustrated as useful for tagging where traditional algorithms have low accuracy.

After these results another finding that is worth mentioning is that the accuracy obtained with the parallel versions of GA, around 97%, is a very

[2] It considers the previous two tags for deciding on the current tag.

**Table 5.4** Comparison of the results of all the algorithms for the two test texts.

| | | GA-Int | | CHC-Bin | | SA-Int | | Viterbi |
|---|---|---|---|---|---|---|---|---|
| Context | | **Best** | **Mean** | **Best** | **Mean** | **Best** | **Mean** | **Best** |
| Brown | 1-0 | 96.01 | 95.26 | 95.32 | 94.91 | 91.83 | 90.95 | **97.04** |
| | 2-0 | 96.41 | 96.23 | 95.35 | 94.80 | 92.28 | 92.14 | **97.48** |
| Susanne | 1-0 | **96.74** | 96.51 | 95.01 | 94.72 | 91.87 | 91.43 | 96.36 |
| | 2-0 | **97.32** | 96.84 | 94.82 | 94.38 | 92.31 | 91.45 | 96.53 |

good result [124] according to the statistical model used. We must take into account that the accuracy is limited by the statistical data provided to the search algorithm. Moreover, the goal of the model is to maximize the probability of the context composed by the tags assigned to a sentence, but it is only an approximate model. The correct tag for a word is not always the most probable one (though most times it is), and the algorithm is conditioned by this fact, but sometimes it is not the one which provides the most probable context either, and it is just in these cases when the tagger fails.

**Table 5.5** Execution times of the different versions of the algorithms (in seconds).

| | | GA-Int | | CHC-Bin | | SA-Int | | Viterbi |
|---|---|---|---|---|---|---|---|---|
| Context | | **Seq.** | **Par.** | **Seq.** | **Par.** | **Seq.** | **Par.** | **Seq.** |
| Brown | 1-0 | 12.31 | 7.02 | 20.48 | 9.60 | 5.84 | 2.98 | 0.47 |
| | 2-0 | 47.93 | 21.19 | 60.92 | 26.44 | 17.32 | 7.93 | 0.47 |
| Susanne | 1-0 | 10.86 | 6.32 | 17.28 | 8.03 | 4.37 | 1.85 | 1.21 |
| | 2-0 | 75.12 | 24.31 | 123.94 | 58.31 | 32.12 | 10.42 | 6.53 |

Let us now analyze Table 5.5, the average execution time for the configurations of the GA, CHC, and SA algorithms, the ones providing the best results for each of them, also including the execution time of the Viterbi method. We can observe that the execution time increases with the size of the context. We can also observe that GA is faster than CHC. Probably, this is due to two reasons: first, binary codings are slower than integer ones, because they require a decodification step prior to apply the fitness function, and second, CHC needs additional computations to detect the converge of the population or to detect incest mating. SA is the fastest of our algorithms. The reason of this is that the SA operates on a single solution, while the rest of the methods are population-based and in addition they execute more complex operators. The table also shows that the parallel implementation reduces the execution time considerably (between 42% and 78%), and this reduction is increasingly beneficial for larger contexts.

If we compare our approach with the Viterbi algorithm we can observe that the classical tagging algorithm is rather fast compared to the rest of our proposed algorithms, though the differences for the Susanne corpus are smaller. Viterbi, and in general any specific algorithm for a problem, will always be more efficient than an evolutionary algorithm, which is a general technique that can be applied to any variant of the problem. However this generality provides some important advantages, such as the applicability to extensions of the model, such as contexts with tags on both sides, in our case. Another advantage concerning the execution time is that there exists parallel versions of the evolutionary algorithm which highly reduce the execution time without requiring addition design effort. In this way the evolutionary algorithm can reach generality and efficiency at the same time.

**Table 5.6** Accuracy and execution times obtained with the GA for the two test texts using two new contexts.

| Context | | Seq | | | Par | | |
|---|---|---|---|---|---|---|---|
| | | **Best** | **Mean** | **Time** | **Best** | **Mean** | **Time** |
| Brown | 1-1 | 96.72 | 96.41 | 56.92 | **96.78** | 96.56 | 19.98 |
| | 2-1 | **96.43** | 96.22 | 210.36 | **96.43** | 96.27 | 67.28 |
| Susanne | 1-1 | 98.36 | 98.11 | 77.14 | **98.59** | 98.39 | 21.25 |
| | 2-1 | 97.78 | 97.31 | 283.49 | **98.01** | 97.64 | 78.17 |

Finally, in order to offer a through analysis, we have also tested our best algorithm (a GA using integer coding) with two more complex contexts (1-1 and 2-1). Table 5.6 shows the results of these experiments. Viterbi can not applied in this case because this algorithm is designed to search the data sequence which maximizes the observed data according to a Markov model, i.e. a model in which the current state only depends on the previous one. If we consider tags on the right of the word being tagged, our model is not a Markov process any more and Viterbi can not be applied. This alone is a strong reason to further research with metaheuristics. As we saw before, the parallelism allows to improve the accuracy and, at the same time, the execution time is reduced considerably (see Table 5.5). In this case, we observe that the increase of the length of the context (from 1-1 to 2-1) provokes a larger execution times.

Though what we propose in this chapter is a search method valid for different tagging models, and thus our goal is not to compete with other models, in order to give an idea of the quality of the particular model that we have used, we present a comparison with the accuracy results of other systems evaluated on the same corpus. This corpus is the Wall Street Journal section of the Penn Treebank [134]. We have used a training set of 554923 words and a test set of 2544 words. Table 5.7 compares our results, obtained with the

GA and the integer representation, with the results obtained by Pla & Molina [135] and Halteren et al. [136]. Pla & Molina proposed a lexicalized HMM taking into account a set of selected words empirically obtained. However, the results shown in Table 5.7 correspond to the non-lexicalized model, because we do not use lexicalization. Halteren et al. propose a combination of different methods for tagging using several voting strategies. The figure which appears in the table for this work corresponds to the results for a single method, which is an HMM. These results confirm the conclusions obtained in the previous experiments. We can observe that in general, our approach has a better behavior when it uses a more complete information, i.e., using a higher context or using tags on the right of the word being tagged. In fact, our general method outperforms the results presented by [135] when right tags are used. Also it is able to find the best known solution [136] using 1-1 contexts.

We can observe that our general approach offers competitive results with respect to the specific methods presented in [135] and [136]. Nevertheless, the quality of our results can be improved by introducing in the model the refinements proposed by the other works, such as lexicalization, combination of different models, etc. Moreover, our method allows extending those models with new features such as right-hand contexts, what can lead to further improvements.

**Table 5.7** Comparison with other systems tagging the Wall Street Journal section of the Penn Treebank.

| Context | **GA-Int** | **Pla & Molina** | **Halteren, Zavrel & Dalelans** |
|---|---|---|---|
| 1-0 | 95.79 | 96.13 | - |
| 2-0 | 96.39 | 96.44 | 96.63 |
| 1-1 | 96.63 | - | - |

The obtained results show that a generic metaheuristic such as our genetic algorithm is able to solve the tagging problem with the same accuracy as an specific method which was designed for this problem. In addition, GAs can perform the search of the best sequence of tags for any context-based model, even if it does not fulfill the Markov assumption. Thus, it is a general method with a proved high quality, and even still able of a later hybridization with problem-dependant operations to yield more accurate results (future line of research).

## 5.5 Conclusions

This chapter compares different optimization methods to solve an important natural language task: the categorization of each word in a text. The optimization methods considered here have been a genetic algorithm (GA), a CHC algorithm, and a simulated annealing (SA). We have compared their results with a widely used method for tagging such as Viterbi.

Results obtained allow extracting a number of conclusions. The first one is that the integer coding performs better than the binary one for the GA and the SA, while the binary one is the best for the CHC algorithm. Parallelism has also proved to be useful, always throwing the more accurate results even with small populations, and reducing the execution time of all the algorithms. The GA has been found to be better than CHC, indicating that the exploration of the search space achieved by the classical genetic operators is powerful enough for this problem. The two evolutionary algorithms have outperformed SA. Also, we have observed that our evolutionary approach is able of outperform classical algorithms such as Viberti in some cases. These results showed that a metaheuristic such as our genetic algorithm is able to solve the tagging problem with the same accuracy as the Viterbi method (an specific method for this problem) with additional scenarios for application forbidden to such specific techniques.

For the future, it could be interesting to investigate other genetic operators for the evolutionary algorithms considered herein, as well as other kinds of metaheuristic methods for tagging.

# 6

# Design of Combinational Logic Circuits

*Computers are useless. They can only give you answers.*

*Pablo Picasso (1881 - 1973) - Spanish painter*

The previous chapter faces a problem related to the Natural Language. To show the versatility of pGAs, we focus now on a very different domain: hardware configuration. In concrete, we apply this metaheuristic to design combinatorial circuits. There are several standard graphical aids widely used by humans to design combinational logic circuits (e.g., Karnaugh Maps [137, 138], and the Quine-McCluskey Method [139, 140]). Despite their advantages, these methods do not guarantee that an optimum circuit can be found given an arbitrary truth table. Additionally, some of these methods (e.g., Karnaugh Maps) have some well-known scalability problems, and can be used only in circuits with very few inputs (normally no more than five or six).

In this chapter, we see the design of combinational logic circuits as an optimization problem in which we aim to find Boolean expressions that produce the outputs required given a set of inputs (as defined by the truth table of a circuit). Seen as an optimization problem, the design of combinational circuits has several interesting features:

- It is a discrete optimization problem in which the decision variables are either integers or binary numbers (as in this chapter). The solutions produced are Boolean expressions that can be graphically depicted.
- The size of the search space grows very rapidly as we increase the number of inputs and/or outputs of a circuit.
- Since it is required to produce circuits that match exactly all the outputs of the truth table given over all the inputs provided, this problem can be considered as having (a usually large number of) hard equality constraints.
- Several parameters of the problem may be modified in order to produce different variations whose degree of difficulty may be higher than that of the original problem. For example, we may vary the types of gates available and the number of inputs that each of them may have.

Because of its complexity, the design of combinational circuits has been tackled with a variety of heuristics (mainly evolutionary algorithms) in the last few years [141, 142]. Despite their good results on small and medium-size

G. Luque and E. Alba: Parallel Genetic Algorithms, SCI 367, pp. 91–114.
springerlink.com 

circuits, heuristics tend to be victims of the "dimensionality curse". Over the years, however, a different goal was envisioned for evolutionary algorithms applied to the solution of combinational logic circuits. The new goal aims to optimize (small and medium-size) circuits (using a certain metric) such that novel designs (since there is no human intervention) can arise. Such novel designs have been shown in the past in a number of studies [141, 142, 143, 144]. In fact, some researchers have pointed out the usefulness of extracting design patterns from such evolutionary-generated solutions. This could lead to a practical design process in which a small (optimal) circuit is used as a building block to produce complex circuits.

This chapter presents a comparative study among a traditional genetic algorithm, simulated annealing, and three heuristics powered by local search capabilities. The rationale behind adopting these approaches is to determine if the design of combinational logic circuits (operating on a binary encoding) can benefit from local search strategies that are not included in a traditional genetic algorithm. For the study, we use both serial and parallel versions of each algorithm, so that we can analyze if the use of parallelism brings any benefits in terms of performance, other than the obvious computational speedup [145].

The remainder of the chapter is organized as follows. In Section 6.1 we provide the statement of the problem of interest to us. In Section 6.2, we briefly discuss the matrix encoding adopted to represent a combinational logic circuit in the heuristics compared. Section 6.3 briefly describes the most relevant previous related work. In Section 6.4, we provide a brief description of the approaches adopted in our study. Section 6.5 contains the examples and the results of the comparative study. Then, there is a further discussion of the results in Section 6.6. Finally, we provide some conclusions and possible ideas of future research in Section 6.7.

## 6.1 Problem Definition

The problem of interest to us consists of designing a circuit that performs a desired function (specified by a truth table), given a certain specified set of available logic gates. This problem is treated, however, as a discrete optimization problem.

In circuit design, it is possible to use various criteria to be minimized. For example, from a mathematical perspective, it is possible to minimize the total number of literals or the total number of binary operations or the total number of symbols in an expression. The minimization problem is difficult for all such cost criteria.

The complexity of a combinational logic circuit is related to the number of gates in the circuit. The complexity of a gate generally is related to the number of inputs to it. Because a logic circuit is a realization (implementation) of a Boolean function in hardware, reducing the number of literals in the function should reduce the number of inputs to each gate and the number of gates in the circuit—thus reducing the complexity of the circuit.

Thus, our overall measure of circuit optimality is the total number of gates used, regardless of their kind. This is approximately proportional to the total part cost of the circuit. Obviously, this minimization criterion is applied only to fully functional circuits (i.e., those that completely match the outputs defined in the corresponding truth table), since it is evidently irrelevant to attempt to minimize infeasible circuits. A *feasible* circuit is one that produces *exactly* all the outputs required for each set of inputs, as indicated in its truth table. To exemplify this, let's consider the Sasao circuit [146]. In this case, we have as a solution the following Boolean expression: $\mathbf{F} = (WX + (Y \oplus W)) \oplus ((X + Y)' + Z)$. So, in order to check feasibility of this circuit, we have to replace each of its inputs (**Z**, **W**, **X** and **Y**) by each of the sets of values depicted in truth table of this circuit [146]. So, in row 1, we have **Z**=0, **W**=0, **X**=0, **Y**=0. By replacing these values in **F** (as defined before), we obtain that **F**=1. This is precisely the value indicated at the end of row 1. Thus, our circuit matches its first output. This same procedure has to be repeated for each of the rows. If the circuit doesn't match any of its required values (e.g., if the output is 1 when it's required to be 0), the circuit is considered to be *infeasible*.

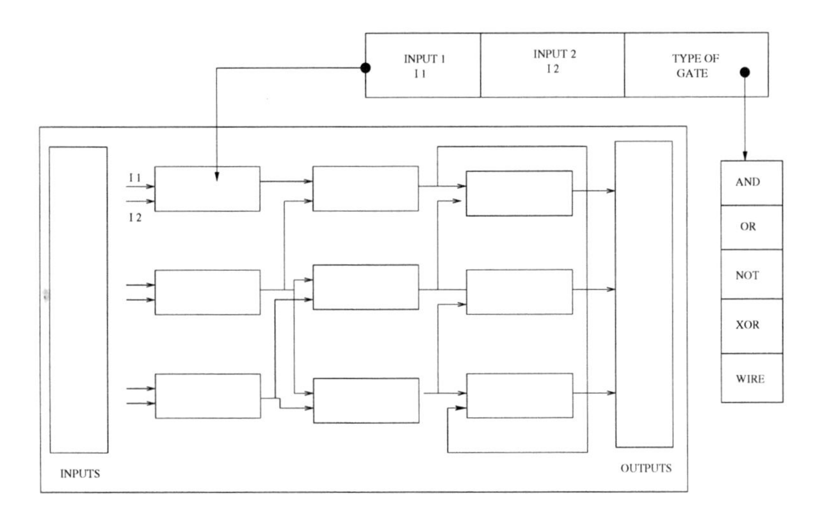


**Fig. 6.1** Matrix used to represent a circuit. Each gate gets its inputs from either of the gates in the previous column. Note the encoding adopted for each element of the matrix as well as the set of available gates used.

Two popular minimization techniques used by electrical engineers are the *Karnaugh Map* [137], which is based on a graphical representation of Boolean functions, and the *Quine-McCluskey Procedure* [139, 140], which is a tabular method. Both of these methods are mechanical in nature. Karnaugh Maps are useful in minimizing the number of literals with up to five or six variables. The Quine-McCluskey Procedure is useful for functions of any number of variables and can easily be programmed to run on a digital computer. Generally, several Boolean function with a minimum number of literals can be obtained for a given truth table using either method, based on the choices made during the minimization process. All minimum functions with the same number of literals yield circuits of the same complexity; hence, any of them can be selected for implementation.

Note that the algebraic simplification process depends entirely on one's familiarity with the postulates and theorems and one's ability to recognize their application. Of course, this ability varies from individual to individual. Depending on the sequence in which the theorems and postulates are applied, more than one simplified form of the expression may be obtained. Usually all such simplified forms are valid and acceptable. Thus, there is (in the general case) no single, unique minimized form of a Boolean expression. However, the solutions that will be shown later on (in Section 6.5) as corresponding to human designers, are really the best solution (based on the minimization of the number of gates, which is our optimization criterion) chosen from a set produced by individuals who can be considered as "expert designers" of combinational logic circuits. Nevertheless, this does not mean that a human cannot improve any of the solutions that we will provide, mainly if we consider that the global optimum of all of the problems adopted remains unknown.

## 6.2 Encoding Solutions into Strings

In order to allow a fair comparison, all of the heuristics compared in this chapter adopted a matrix to represent a circuit as in our previous work [141, 147] (see Fig. 6.1).

More formally, we can say that any circuit can be represented as a bidimensional array of gates $S_{i,j}$, where $j$ indicates the *level* of a gate, so that those gates closer to the inputs have lower values of $j$. (Level values are incremented from left to right in Fig. 6.1). For a fixed $j$, the index $i$ varies with respect to the gates that are "next" to each other in the circuit, but without being necessarily connected. Each matrix element is a gate (there are 5 types of gates: AND, NOT, OR, XOR and WIRE[1]) that receives its 2 inputs from any gate at the previous column as shown in Fig. 6.1. It is important to clarify that the number of rows and columns of the matrix used to encode a circuit are values defined by the user. Given a circuit to be optimized, we suggest to use the following procedure in order to define the matrix size (i.e., number of rows and columns) to encode it:

[1] WIRE basically indicates a null operation, or in other words, the absence of gate, and it is used just to keep regularity in the representation used. Otherwise, we would have to use variable-length strings.

1. Start with a square matrix of size 5 (i.e., number of rows = number of columns = 5).
2. If no feasible solution is found using this matrix, then increase the number of columns by one, without changing the number of rows.
3. If no feasible solution is found using this matrix, then increase the number of rows by one, without changing the number of columns.
4. Repeat steps 2 and 3 until a suitable matrix is produced. In each case, at least 10 independent runs (using different random seeds for the initial population) must be performed in order to determine feasibility. If none of these runs produces at least one feasible solution, then it is considered that "no feasible solution was found".

As we will see in Table 6.1 from Section 6.5, it is normally the case that for small circuits a matrix of $5 \times 5$ is sufficient. However, in two of our case studies reported in Section 6.5, we reached a matrix size of $6 \times 7$. This situation normally arises with circuits having several outputs, although in some cases, such as in the 2-bit multiplier described in Section 6.5, even a $5 \times 5$ matrix is enough to find the best-known circuit. The above guidelines have been successfully adopted with a variety of circuits in some of our previous work [141].

A chromosomic string encodes the matrix shown in Fig. 6.1 by using triplets in which the 2 first elements refer to each of the inputs used, and the third is the corresponding gate from the available set.

The matrix representation adopted in this work was originally proposed by Louis [148, 149]. He applied his approach to a 2-bit adder and to the $n$-parity check problem (for $n = 4, 5, 6$). This representation has also been adopted by Miller et al. [142, 150] in the so-called *Cartesian Genetic Programming* with some differences. For example, the restrictions regarding the source of a certain input to be fed in a matrix element varies in each of the three approaches: Louis [148] has strong restrictions, Miller et al. [150] have no restrictions and we have relatively light restrictions. Although our representation allows the case with no restrictions, we decided to keep its original restrictions as to allow a fair comparison with some of our previous work.

It is worth emphasizing that the use of matrix-based encodings such as the one adopted here results particularly useful for designing combinational logic circuits, since they do not allow *bloat* (i.e., the uncontrolled tree growth normally associated with traditional genetic programming [151]) [141, 142].

The following formula is used to compute the fitness of an individual $\mathbf{x}$ for all the heuristics compared in this chapter:

$$fitness(\mathbf{x}) = \begin{cases} \sum_{j=1}^{p} f_j(\mathbf{x}) & \text{if } f(\mathbf{x}) \text{ is not feasible} \\ \sum_{j=1}^{p} f_j(\mathbf{x}) + w(\mathbf{x}) & \text{otherwise} \end{cases} \tag{6.1}$$

where $p$ is the number of entries of the truth table (normally, $p = 2^n$, being $n$ the number of inputs of the truth table, but $p$ can also be assigned a certain value directly, in case the truth table has "don't cares"), and the value of $f_j(\mathbf{x})$ depends on the outcomes produced by the circuit $\mathbf{x}$ encoded (whenever the solution produced matches the corresponding entry of the truth table at location $j$, a value of one is assigned to $f_j(\mathbf{x})$; otherwise, a value of zero is assigned). The function $w(\mathbf{x})$ returns an integer equal to the number of WIREs present in the circuit $\mathbf{x}$ encoded. The solutions produced are Boolean expressions which will be made of Boolean operators (AND, OR, NOT, XOR) and of variables which take only binary values (either zero or one). The solutions (i.e., the circuits obtained) will be expressed in two forms: (1) through its Boolean expression(s) and (2) by showing its graphical representation. In order to understand both, the Boolean expressions and the graphical representations of the circuits, the reader must rely on the symbols shown in Fig. 6.2. Note that the AND operator is assumed by default in the Boolean expressions. Thus, **AB** must be interpreted as: **A** AND **B**.

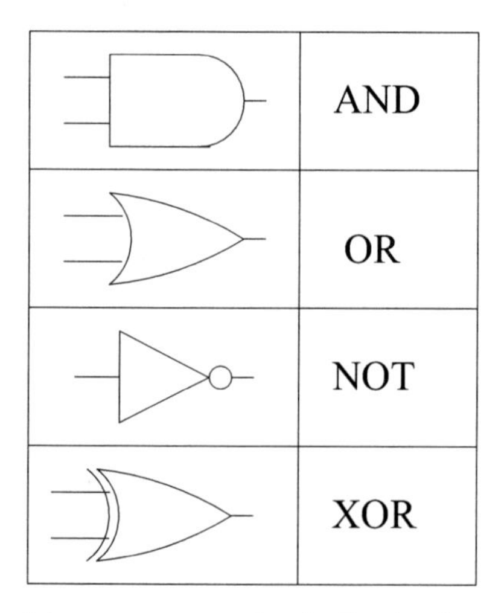


**Fig. 6.2** Symbols used to represent a circuit. In the first column, we show the graphical symbol for each gate. In the second column, we show the mathematical symbol adopted in the Boolean expressions. In the third column, we show the name of each of the Boolean operators adopted.

In words, we can say that our fitness function works in two stages [141]: first, it maximizes the number of matches (as in Louis' case). However, once feasible solutions (i.e., the circuit generated by the solution computes the objective truth table) are found, we maximize the number of WIREs in the circuit. By doing this, we actually optimize the circuit in terms of the number of gates that it uses.

Thus, we can say that our goal is to produce a fully functional design (i.e., one that produces all the expected outputs for any combination of inputs according to the truth table given for the problem) which maximizes the number of WIREs.

## 6.3 Related Works

Despite the considerable amount of work currently available on the use of genetic algorithms, genetic programming and evolution strategies to design combinational logic circuits in the last few years (see for example [141, 142, 148]), there have been few attempts to compare different heuristics in this problem. Here, the main motivation for such a comparative study is to analyze whether certain types of heuristics (namely, hybrid approaches and local search methods) could be more suitable for this type of problems than the use of traditional genetic algorithms.

Previous work has found, among other things, that designing combinational logic circuits is highly sensitive to the encoding [148, 152, 153], and to the degree of interconectivity allowed among gates [154]. There have also been studies on the fitness landscapes of these problems that finally rate the problem as being quite difficult for an evolutionary algorithm [155, 156]. However, this sort of analysis has been conducted only on a single type of heuristics (e.g., a genetic algorithm [141], an evolution strategy [142], simulated evolution [152], the ant colony [157, 158], or particle swarm optimization [159, 160]). Additionally, given the scalability problem associated with the design of combinational logic circuits using evolutionary algorithms, the use of parallelism seems a capital issue [161]. Remarkably, however, few studies available in the literature have considered parallelism in the past. Thus, we also consider in this chapter the use of parallel versions of the algorithms compared as to analyze the way in which parallelization affects the exploration of the search space in the specific domain of our interest.

## 6.4 Sequential, Parallel, and Hybrid Approaches

In this chapter, we compare five heuristics for the design of circuits:

1. A genetic algorithm (GA) with binary representation such as the one described in [141, 147]. The main motivation for using this approach was our previous experience (and relative success) applying this heuristics to design combinational logic circuits [141]. Genetic Algorithms (GAs) [4, 162] are stochastic search methods that have been successfully applied in many real applications of high complexity. A GA is an iterative technique that applies stochastic operators on a pool of individuals (tentative solutions). An evaluation function associates a value to every individual indicating its suitability to the problem. A GA usually applies a recombination operator on two solutions, plus a mutation operator that randomly modifies the individual contents to promote diversity. In our experiments we use the uniform crossover (UX) and the Bit-Flip mutation. The UX consists in creating two offspring with each allele in the new offspring taken randomly from one parent. The Bit-Flip mutation works by probabilistically changing every position (allele) to its complementary value. For full details

about this operators see [4, 35, 130]. The pseudo-code of the GA adopted is shown in Fig. 6.3. In all the pseudo-codes, the evaluation phase represents that the fitness function (Eq. 1) is evaluated on the respective population.

```
1 t = 0
2 initialize P(t)
3 evaluate structures in P(t)
4 while not end do
5       t     = t + 1
6       C(t)  = selectFrom(P(t-1))
7       C'(t) = recombine(C(t))
8       C'(t) = mutate(C'(t))
9       evaluate structures in C'(t)
10      replace P(t) from C'(t) and P(t-1)
11 endwhile
12 return best found solution
```

**Fig. 6.3** Scheme of the GA adopted.

2. A CHC [131] which is a variant of the genetic algorithm with a particular way of promoting diversity. It uses a highly disruptive crossover operator to produce new individuals maximally different from their parents. It is combined with a conservative selection strategy which introduces a kind of inherent elitism. Fig. 6.4 shows a scheme of the CHC algorithm, whose main features are:
   - The mating is not restricted to the best individuals, but parents are randomly paired in a mating pool $C(t)$ (line 6 of Fig. 6.4). However, recombination is only applied if the Hamming distance between the parents is above a certain threshold, a mechanism of *incest prevention* (line 8 of Fig. 6.4).
   - CHC uses a *half-uniform crossover* (HUX), which exchanges exactly half of the differing parental genes (line 9 of Fig. 6.4). HUX guarantees that the children are always at the maximum Hamming distance from their two parents.
   - Traditional selection methods do not guarantee the survival of best individuals, though they have a higher probability to survive. On the contrary, CHC guarantees survival of the best individuals selected from the set of parents ($P(t-1)$) and offspring ($C'(t)$) put together (line 11 of Fig. 6.4).
   - Mutation is not applied directly as an operator.
   - CHC applies a re-start mechanism if the population remains unchanged for some number of generations (lines 12-13 of Fig. 6.4). The new population includes one copy of the best individual, while the rest of the population is generated by mutating some percentage of bits of such

best individual. The main motivation for using CHC was to see if the use of a highly disruptive crossover operator would have a positive effect on a genetic algorithm when optimizing combinational circuits.

```
1 t = 0
2 initialize P(t)
3 evaluate structures in P(t)
4 while not end do
5       t = t + 1
6       select: C(t) = P(t-1)
7       for each pair (p1,p2) in C(t)
8           if 'incest prevention condition'
9               add to C'(t) HUX(p1,p2)
10      evaluate structures in C'(t)
11      replace P(t) from C'(t) and P(t-1)
12      if convergence(P(t))
13          re-start P(t)
14 endwhile
15 return best found solution
```

**Fig. 6.4** Scheme of the CHC algorithm.

3. A simulated annealing (SA) algorithm. The simulated annealing algorithm was first proposed in 1983 [12] based on a mathematical model originated in the mid-1950s. SA [163, 164] is a stochastic relaxation technique that can be seen as a hill-climber with an internal mechanism to escape local optima. It is based in a cooling procedure used in the metallurgical industry. This procedure heats the material to a high temperature so that it becomes a liquid and the atoms can move relatively freely. The temperature is then slowly lowered so that at each temperature the atoms can move enough to begin adopting the most stable configuration. In principle, if the material is cooled slowly enough, the atoms are able to reach the most stable (optimum) configuration. This smooth cooling process is known as *annealing*. Fig. 6.5 shows a scheme of SA. First at all, the parameter $T$, called the temperature, and the solution, are initialized (lines 2-4). The solution $s1$ is accepted as the new current solution if $\delta = f(s1) - f(s0) > 0$. Stagnations in local optima are prevented by accepting also solutions which increase the objective function value with a probability $exp(\delta/T)$ if $\delta < 0$. This process is repeated several times to obtain good sampling statistics for the current temperature. The number of such iterations is given by the parameter $Markov\_Chain\_length$, whose name alludes the fact that the sequence of accepted solutions is a Markov chain (a sequence of states in which each state only depends on the previous one). Then the temperature is decremented (line 14) and the entire process is repeated until a

frozen state is achieved at $T_{min}$ (line 15). The value of $T$ usually varies from a relatively large value to a small value close to zero. Here, we are using the Fast SA scheme ($T_k = T_0/(1+k)$) for updating the temperature. Considering the well-known success of simulated annealing in a variety of optimization problems (both on combinatorial and on continuous search spaces), the main motivation to adopt it in this problem was clearly to see if its local search capabilities would be better than the global search capabilities of a genetic algorithm in the design of combinational logic circuits.

```
1  t = 0
2  initialize(T)
3  s0 = Initial_Solution()
4  v0 = Evaluate(s0)
5  repeat
6      repeat
7          t  = t + 1
8          s1 = Generate(s0,T)
9          v1 = Evaluate(s0,T)
10         if Accept(v0,v1,T)
11             s0 = s1
12             v0 = v1
13     until t mod Markov_Chain_length == 0
14     T = Update(T)
15 until 'loop stop criterion' satisfied
16 return best found solution
```

**Fig. 6.5** Scheme of the Simulated Annealing (SA) algorithm.

4. Finally, we define two hybrid algorithms. In its broadest sense, hybridization refers to the inclusion of problem-dependent knowledge in a general search algorithm [130] in one of two ways: *strong hybrids*, where problem-knowledge is included as problem-dependent representation and/or operators, and *weak hybrids*, where several algorithms are combined in some manner to yield the new hybrid algorithm. First, we define a weak hybrid called GASA1, where a GA uses SA as an evolutionary operator. The figure and the pseudo-code of this approach is shown in Fig. 6.6. In the main loop of this method after the traditional recombination and mutation operators are applied (lines 7 and 8), several solutions are randomly selected (according to a low probability) from the current offspring and they are improved using the local search algorithm (line 9). The rationale for this sort of hybridization is that, while the GA locates "good" regions of the search space (exploration), SA allows for exploitation in the best regions found by its partner. Evidently, the motivation in this case was to see if by taking the best of these two heuristics (i.e., the genetic algorithm

and simulated annealing), we could produce another heuristic which would perform better than any of the two approaches from which it was created.

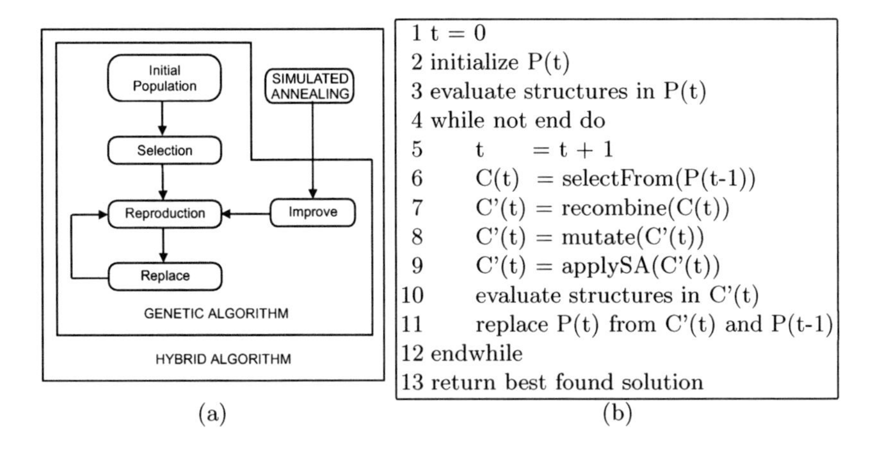


**Fig. 6.6** Model of Hybridization 1 (GASA1).

5. A second weak hybrid scheme called GASA2, which executes a GASA1 until the algorithm completely finishes. Then the hybrid selects (by tournament [4]) some individuals from the final population and starts a SA algorithm over them. The main motivation for this approach was to see if simulated annealing could use its local search capabilities to improve solutions generated by another approach, and which would presumably be close to the global optimum. The pseudo-code of this approach is shown in Fig. 6.7.

As we described in Chapter 2.5, a parallel EA (PEA) is an algorithm having multiple component EAs, regardless of their population structure. In this chapter, we have chosen a kind of decentralized distributed search. In this parallel implementation separate subpopulations evolve independently in a ring with sparse asynchronous exchanges of one individual with a certain given frequency. The selection of the emigrant is through binary tournament in the evolutionary algorithms, and the arriving immigrant replaces the worst one in the population only if the new one is better than this current worst individual.

For the parallel SA there also exist multiple asynchronous component SAs. Each component SA periodically exchanges the best solution found (cooperation phase) with its neighbor SA in the ring.

Although many other hybrid approaches for optimization exist (see for example [165, 166, 167, 168, 169]), we decided to adopt only the approaches previously described because the optimization problem of our interest is discrete, subject to a (usually large) set of equality constraints and in which the decision variables are actually binary numbers. Most of the hybrids (particularly those involving simulated annealing) that we found in the literature have been applied either to combinatorial optimization problems (in which the decision variables are permutations of integers), or to global optimization problems (in which the decision variables are real numbers). In fact, although many heuristics have been applied to the design of combinational logic circuits (e.g., [141, 142, 152, 157, 159, 160]), no hybrid approach has been previously adopted in this problem, to the authors' best knowledge, mainly because of the peculiar features of this problem (when seen as an optimization task). As previously discussed, the approaches adopted for our comparative study were carefully designed to tackle the problem of our interest. However, this is not to say that these are the only approaches that can be applied to the design of circuits, since many other heuristics and many other hybrids may be designed for that purpose.

## 6.5 Computational Experiments and Analysis of Their Results

We compare our binary GA with respect to SA, CHC, GASA1 and GASA2 both in serial and parallel versions. In Table 6.1 we summarize the features of the problem instances that we use in our experiments.

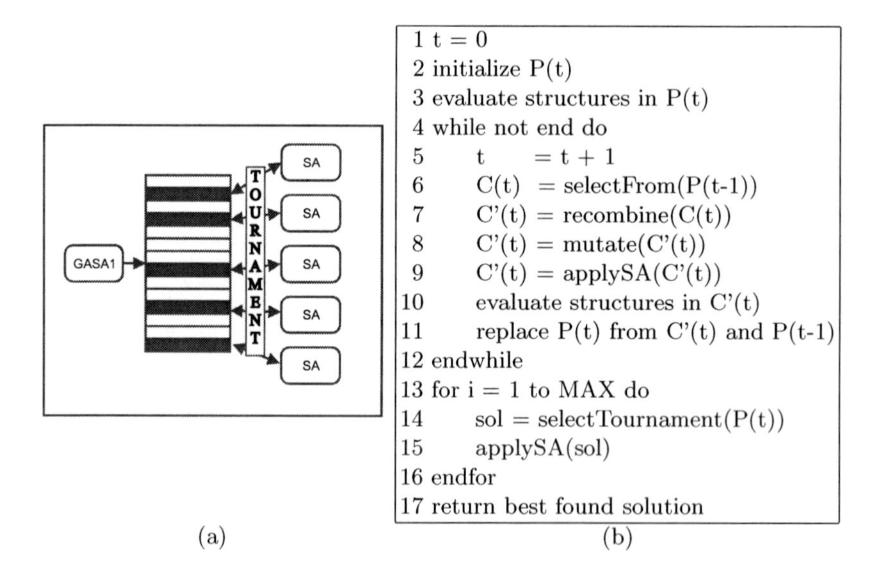


**Fig. 6.7** Model of Hybridization 2 (GASA2).

**Table 6.1** Features of the circuits. **size** = matrix size in rows × columns, **codesize** = length of the binary string, **BKS** = best-known solution (i.e., the fitness value of the best solution reported in the literature for the corresponding circuit).

| name | inputs | outputs | size | codesize | BKS |
|---|---|---|---|---|---|
| Sasao | 4 | 1 | $5 \times 5$ | 225 | 34 [146] |
| Catherine | 5 | 1 | $6 \times 7$ | 278 | 67 [170] |
| Katz 1 | 4 | 3 | $6 \times 7$ | 278 | 81 [146] |
| 2-bit multiplier | 4 | 4 | $5 \times 5$ | 225 | 82 [146] |
| Katz 2 | 5 | 3 | $5 \times 5$ | 225 | 114 [171] |

Since our main goal was to analyze the behavior of different heuristics and the impact of parallelism, no particular effort was placed in fine-tuning the parameters for each of the circuits tried. The population sizes, mutation, and crossover rates used correspond to the values previously reported for a traditional (binary) GA [147]. In all the evolutionary algorithms, the population is composed of 320 individuals for the first case study, while 600 individuals are used for the other four. All experiments use a crossover rate of 60% and a mutation rate of 50% of the chromosome length. The CHC method restarts the population (an uniform mutation ($p_m = 0.7$) is applied to the 35% of the population) whenever convergence is detected. The hybrid GASA1 uses the SA operator (100 iterations for the first and third case studies and 500 iterations for the rest) with probability 0.01, i.e., this improvement process only is applied to approximately one of each 100 solutions of the current offspring. The second hybrid (GASA2) executes a SA (with 3000 iterations for the first instance and 10000 for the rest) when GASA1 finishes. The migration in dEAs occurs in a unidirectional ring manner, sending one single individual (chosen by binary tournament) to the neighboring sub-population. The target population incorporates this individual only if it is better than its presently worst solution. The migration step is performed every 20 iterations in every island in an asynchronous way. The selected migration policy configuration allows to maintain a global good diversity, and to lead the global search to good regions of the search space. The asynchronous communications that we used provokes that the communication overhead was insignificant. Since we want to compare against the sequential EAs, dEAs use the same population size, but now the whole population of the sequential EA is split into as many subpopulations as processes involved in the parallel computation. Our parallel algorithms are composed of eight subpopulations. Finally, the number of iterations of the SA has been chosen in order to compute a similar number of evaluations as to the GA, and the Markov chain length is preset to $max\_iter/10$. We performed 20 independent runs per algorithm per circuit per version (either serial or parallel) using the parameters summarized above.

The most relevant aspects that were measured in this comparison are the following ones: best fitness value obtained (we call this **opt**), the number of times that the approach found the best fitness value (we call this **hits**), the average final fitness (called **avg**), and the average number of fitness function evaluations required to find the best fitness value reported (**#evals**).

A short note regarding the stopping criteria adopted is in place. Each algorithm stops when reaching the target fitness or a maximum (predefined) number of generations. At the end of each generation, the algorithm checks if the stopping criterion is satisfied, i.e., if the current generation number exceeds the predefined limit or if an end signal has been received (for parallel executions).

### *6.5.1 Case Study 1: Sasao*

Our first case study has 4 inputs and one output. Our comparison of results for this case study is shown in Table 6.2. In this case both GASA1 and GASA2 were able to converge to the best known solution for this circuit (which has 7 gates and a fitness of 34) [146]. The best solution found is: F = $(WX + (Y \oplus W)) \oplus ((X + Y)' + Z)$. Note that both, GASA1 and GASA2, required the highest number of evaluations to reach their best fitness value, but their final solution was significantly better than the solutions found by the other algorithms. Also note that the parallel versions of GASA1 and

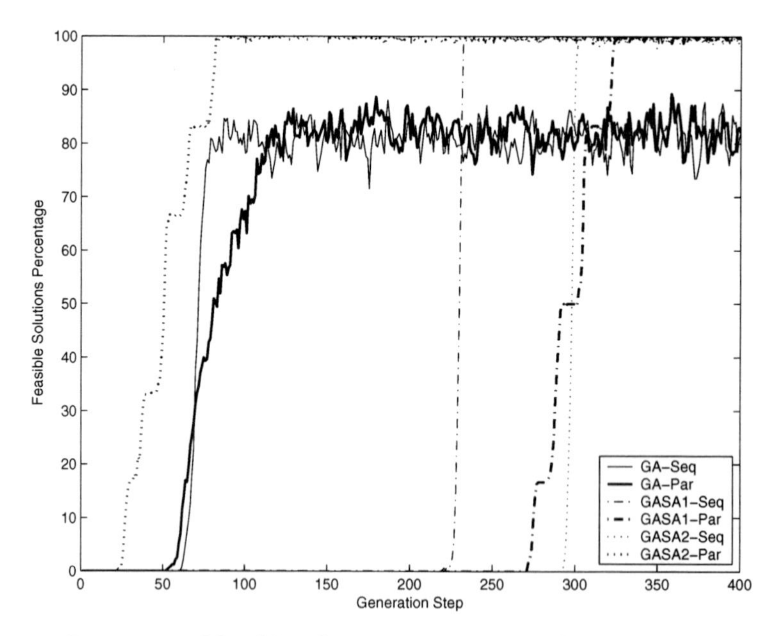


**Fig. 6.8** Percentage of feasible solutions per generation for the circuit of the first case study.

GASA2 increased the average fitness value and the number of hits. However, the average number of fitness function evaluations to find the best fitness value did not decrease in the parallel versions of GASA1 and GASA2, as it occurred for the parallel versions of the traditional GA, CHC, and SA. Finally, we observed that the average fitness value of parallel SA was slightly worse than the value of the serial version, which indicates that the parallel algorithm behavior is not adequate for this instance. Interestingly, SA was the only approach whose average fitness did not increase when using parallelism.

**Table 6.2** Comparison of results for the first case study.

| **Algorithm** | **sequential** | | | | **parallel** | | | |
|---|---|---|---|---|---|---|---|---|
| | **opt** | **hits** | **avg** | **#evals** | **opt** | **hits** | **avg** | **#evals** |
| GA | 31 | 10% | 15.8 | 96806 | 33 | 5% | 18.1 | 79107 |
| CHC | 27 | 5% | 15.1 | 107680 | 32 | 5% | 16.4 | 75804 |
| SA | 30 | 35% | 15.6 | 70724 | 31 | 5% | 15.2 | 69652 |
| GASA1 | **34** | 10% | 23.2 | 145121 | **34** | 20% | 25.5 | 151327 |
| GASA2 | **34** | 10% | 24.2 | 147381 | **34** | **30%** | 27.8 | 155293 |

Another aspect that is worth analyzing is the percentage of feasible solutions that each algorithm maintains along the evolutionary process. Such a percentage gives an idea of how difficult is for each approach to reach the feasible region and to maintain feasibility. Fig. 6.8 shows the (average) percentage of feasible solutions present in the population over time (i.e., generations) for each of the algorithms compared. It is particularly interesting to note how the parallel version of GASA2 starts increasing its percentage of feasible solutions rather quickly and reaches 100% feasibility in less than 100 generations. It is also worth commenting on the GA, which was never able to reach a feasibility rate of 100% (in any of its two versions). All the other approaches were able to reach 100% feasibility, but much later than the parallel version of GASA2. Thus, we can conclude that, in this case study, GASA2 was the best overall solver in its two versions. GASA2 produced the highest average fitness, the highest number of hits and was the fastest to reach the feasible region and to reach 100% feasibility.

**Table 6.3** Comparison of the best solutions found for the first case study by GASA2, the $n$-cardinality genetic algorithm (**NGA**) [147], a human designer (**HD 1**) who used Karnaugh maps and theorems from Boolean algebra, and **Sasao** [172], who used this circuit to illustrate his circuit simplification technique based on the use of ANDs & XORs.

| **GASA2** | **NGA** | **HD 1** | **Sasao** |
|---|---|---|---|
| 7 gates | 10 gates | 11 gates | 12 gates |

Just to give an idea on how good is the solution found by GASA2, we show in Table 6.3 a comparison of the best solution found by GASA2 with respect to existing approaches for the first problem. This second comparison is only in terms of the Boolean expression found. Note that the $n$-cardinality GA (NGA) used the same parameters as its binary counterpart. We can see that GASA2 found a solution significantly better than the other approaches with respect to which it was compared (the $n$-cardinality GA, Sasao's simplification technique based on the use of ANDs & XORs [172], and a human designer using Karnaugh maps).

### *6.5.2 Case Study 2: Catherine*

Our second case study has 5 inputs and one output and our comparison of results is shown in Table 6.4. Again, GASA2 found the best solution, but in this case, the parallel version produced a slightly better result (**opt** column) than its serial counterpart. The best solution found for this case study is: F $= ((A_4)'(A_2A_0 + A_1)(A_2 + A_0))'((A_2A_0 + A_1)(A_2 + A_0) + A_3)$. Note also that the average fitness was increased both for GASA1 and GASA2 in their parallel versions. Furthermore, it is worth noticing that in this case the use of parallelism decreased the average number of evaluations required to find the best possible fitness value produced by each of the algorithms under study. Except for CHC, all the other approaches improved their average fitness when using parallelism. Another important detail is that the sequential SA outperformed the GA in locating a larger final best fitness value with a significant reduction in evaluations, although the SA obtained a worse average fitness than the GA.

**Table 6.4** Comparison of results for the second case study.

| **Algorithm** | **sequential** | | | | **parallel** | | | |
|---|---|---|---|---|---|---|---|---|
| | **opt** | **hits** | **avg** | **#evals** | **opt** | **hits** | **avg** | **#evals** |
| GA | 60 | 5% | 36.5 | 432170 | 62 | 10% | 41.0 | 345578 |
| CHC | 58 | 15% | 29.8 | 312482 | 61 | 5% | 28.9 | 246090 |
| SA | 61 | 5% | 33.1 | 175633 | 62 | 5% | 34.2 | 154064 |
| GASA1 | 63 | 40% | 45.1 | 694897 | **65** | 5% | 50.6 | 593517 |
| GASA2 | 64 | 10% | 47.3 | 720106 | **65** | **10%** | 52.9 | 609485 |

Fig. 6.9 shows the (average) percentage of feasible solutions present in the population over time (i.e., generations) for each of the algorithms compared. Again, the parallel version of GASA2 starts increasing its percentage of feasible solutions rather quickly. In this case, it reaches 100% feasibility in less than 300 generations. The second best performer in this case was the parallel version of GASA1, reaching 100% feasibility in about 500 generations. The sequential version of the GA was the only approach unable to reach 100% feasibility.

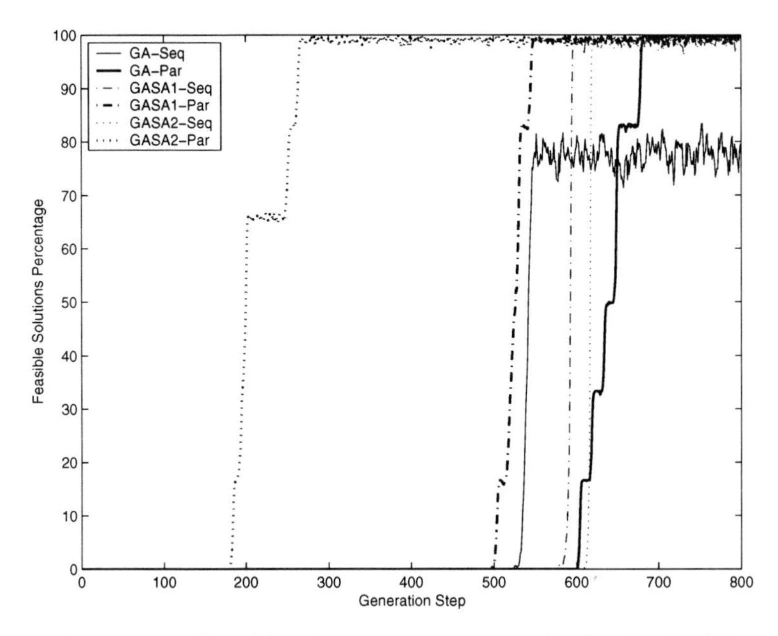


**Fig. 6.9** Percentage of feasible solutions per generation for the circuit of the second case study.

**Table 6.5** Comparison of the best solutions found for the second case study by GASA2, the $n$-cardinality genetic algorithm (**NGA**) [147], and a human designer (**HD 1**) who used Karnaugh maps and theorems from Boolean algebra.

| **GASA2** | **NGA** | **HD 1** |
|---|---|---|
| 9 gates | 10 gates | 12 gates |

To give an idea on how good is the solution found by GASA2, we show in Table 6.5 a comparison of the best solution found by GASA2 with respect to the solutions found by other approaches. GASA2 improved the best solution found both by the NGA and by a human designer (using Karnaugh maps). Clearly, GASA2 was the best overall performer in this case study as well. It is also important to mention that the best solution found by GASA2, which has 9 gates, is not the best possible solution for this circuit (there is another one with only 7 gates: $F = (A + BC)(D \oplus E)(B + C) \oplus D)$, which has been obtained with genetic programming [170]. However, as indicated before, no attempt was made to fine-tune the parameters of the algorithms used as to achieve a better solution.

### 6.5.3 Case Study 3: Katz 1

Our third case study has 4 inputs and 3 outputs and the comparison of results is shown in Table 6.6. In this case, both GASA1 and GASA2 found the best solution reported in the literature for this circuit [146], which has 9 gates and fitness 81. However, note that GASA2 had a better hit rate (in the parallel version). The best solution found is: $F_1 = ((D \oplus B) + (A \oplus C))'$, $F_2 = ((D \oplus B) + (A \oplus C))(C \oplus ((A \oplus C) + (A \oplus B))) \oplus ((D \oplus B) + (A \oplus C))$, $F_3 = (C \oplus ((A \oplus C) + (A \oplus B)))((D \oplus B) + (A \oplus C))$. In this case, the use of parallelism produced a noticeable increment in the average fitness of GASA1 and GASA2, but the best solution was only rarely found. It is also interesting to see how GASA1 and GASA2 both have a computational cost of twice that of the traditional GA. Also note that, as in the previous case studies, in this one the use of parallelism decreased the average number of evaluations required to find the best possible fitness value produced by each of the algorithms under study. All the approaches improved their average fitness when using parallelism. The behavior of parallel SA was slightly different to the rest of algorithms for this instance, always showing a very small number of evaluations at the price of a medium-low hit rate.

**Table 6.6** Comparison of results for the third case study.

| **Algorithm** | **sequential** | | | | **parallel** | | | |
|---|---|---|---|---|---|---|---|---|
| | **opt** | **hits** | **avg** | **#evals** | **opt** | **hits** | **avg** | **#evals** |
| GA | 71 | 10% | 51.2 | 552486 | 76 | 15% | 54.5 | 498512 |
| CHC | 64 | 20% | 47.3 | 362745 | 70 | 5% | 49.3 | 252969 |
| SA | 67 | 15% | 46.3 | 194573 | 71 | 5% | 51.3 | 197315 |
| GASA1 | 78 | 35% | 70.0 | 1090472 | **81** | 5% | 76.1 | 963482 |
| GASA2 | 78 | 5% | 69.3 | 1143853 | **81** | **10%** | 77.9 | 1009713 |

Fig. 6.10 shows the (average) percentage of feasible solutions present in the population over time (i.e., generations) for each of the algorithms compared. Interestingly, the sequential version of the GA was the approach that reached the feasible region more quickly in this case study, being able to reach 100% feasibility before generation 200. The sequential version of GASA2 was the second best performer. However, all the approaches were able to reach 100% feasibility before generation 500, which is an indicative of the fact that the search space of this problem is not as rough as that of the previous case studies. Once more, GASA2 can be considered the best overall performer, since it produced the highest average fitness and was able to reach more consistently (in its parallel version) the best-known solution for this case study.

When performing a comparison of these results with respect to other approaches (Table 6.7), it is worth indicating that GASA2 again improved on the best solution found by two human designers (one using Karnaugh maps and the other one using the Quine-McCluskey method), and by the NGA.

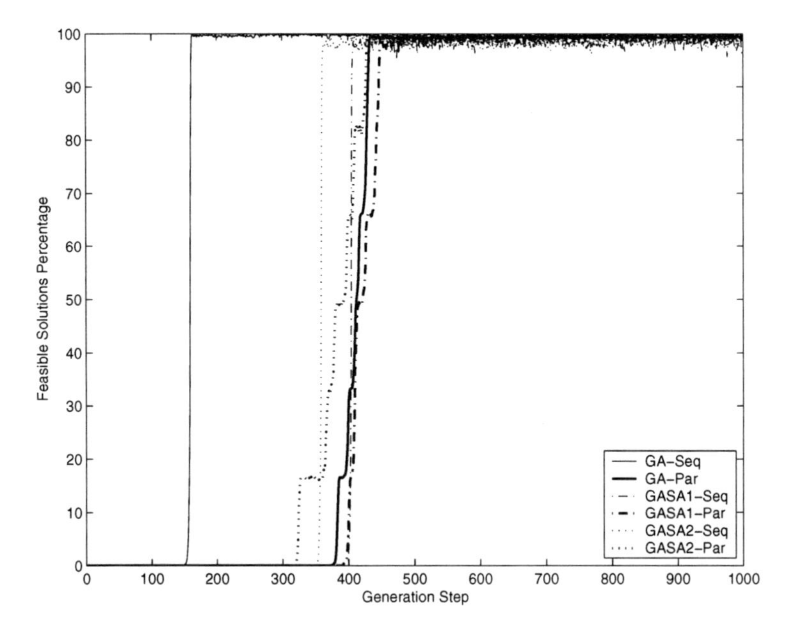


**Fig. 6.10** Percentage of feasible solutions per generation for the circuit of the third case study.

**Table 6.7** Comparison of the best solutions found for the second case study by GASA2, the $n$-cardinality genetic algorithm (**NGA**) [147], a human designer (**HD 1**) who used Karnaugh maps and theorems from Boolean algebra and a second human designer (**HD 2**) who used the Quine-McCluskey method.

| GASA2 | NGA | HD 1 | HD 2 |
|---|---|---|---|
| 9 gates | 12 gates | 19 gates | 13 gates |

## *6.5.4 Case Study 4: 2-Bit Multiplier*

Our fourth case study has 4 inputs and 4 outputs and the results are shown in Table 6.8. In this case, GASA2 found the best solution reported in the literature for this circuit [146], which has 7 gates and a fitness value of 82. The best solution found for this case study is: $C_3 = (B_0A_1)(B_1A_0)$, $C_2 = (A_1B_1) \oplus (B_0A_1)(B_1A_0)$, $C_1 = (B_0A_1) \oplus (B_1A_0)$, $C_0 = A_0B_0$. The use of parallelism for this instance produced only a slight increase in the average fitness of GASA1 and GASA2, but allowed GASA2 to converge to the best solution reported in the literature. In fact, all the approaches improved their average fitness when using parallelism. It is also interesting to see how GASA1 and GASA2 both have a computational cost much higher than the traditional GA. Note however, that the parallel version of the parallel GA was able to converge to a better solution than the parallel version of GASA1, although the average fitness of the GA was still slightly below GASA1. The GA obtained better results (**opt** and **avg** columns) than the other pure algorithms (SA and CHC), but it required a higher number of evaluations.

**Table 6.8** Comparison of results for the fourth case study.

| Algorithm | sequential | | | | parallel | | | |
|---|---|---|---|---|---|---|---|---|
| | **opt** | **hits** | **avg** | **#evals** | **opt** | **hits** | **avg** | **#evals** |
| GA | 78 | 15% | 71.8 | 528390 | 81 | 5% | 76.3 | 425100 |
| CHC | 76 | 5% | 72.7 | 417930 | 80 | 10% | 74.2 | 246090 |
| SA | 77 | 5% | 68.6 | 268954 | 77 | 10% | 69.3 | 234562 |
| GASA1 | 78 | 25% | 74.1 | 711675 | 80 | 20% | 76.9 | 852120 |
| GASA2 | 80 | 10% | 75.4 | 817245 | **82** | **20%** | 78.7 | 927845 |

Fig. 6.11 shows the (average) percentage of feasible solutions present in the population over time (i.e., generations) for each of the algorithms compared. In this case, the parallel version of GASA2 was, once more, the fastest approach to reach both the feasible region and a 100% feasibility (this was achieved before generation 200). The second best performer in terms of feasibility was the sequential version of the GA, which reached 100% by generation 400. However, its sequential counterpart was the worst performer. Note also that in this case study, all the approaches were able to reach 100% feasibility. GASA2 was again the best overall performer. In its parallel version, GASA2 was the only approach able to reach the best-known solution for this case study.

We show in Table 6.9 a comparison of the best solution found by GASA2 with respect to other approaches previously used to design the circuit of the fourth case study. This second comparison is only in terms of the Boolean expression found. In this case, GASA2 again improved on the best solution found by two human designers (one using Karnaugh maps and the other one using the Quine-McCluskey method), by the NGA and by the cartesian genetic programming of [150]. It should be mentioned that Miller et al. [150] considered their solution to contain only 7 gates because of the way in which they encoded their Boolean functions (the reason is that they encoded NAND gates in their representation). However, since we considered each gate as a separate chromosomic element, we count each of them, including NOTs that are associated with AND & OR gates. It is also worth noticing that Miller et al. [150] found their solution with runs of 3,000,000 fitness function evaluations each, much longer than ours (inefficiently).

### *6.5.5 Case Study 5: Katz 2*

Our fifth case study has 5 inputs and 3 outputs. Note that despite the size of the truth table, a $5 \times 5$ matrix was also adopted in this case. Our comparison of results is shown in Table 6.10. In this case, both GASA1 and GASA2 found

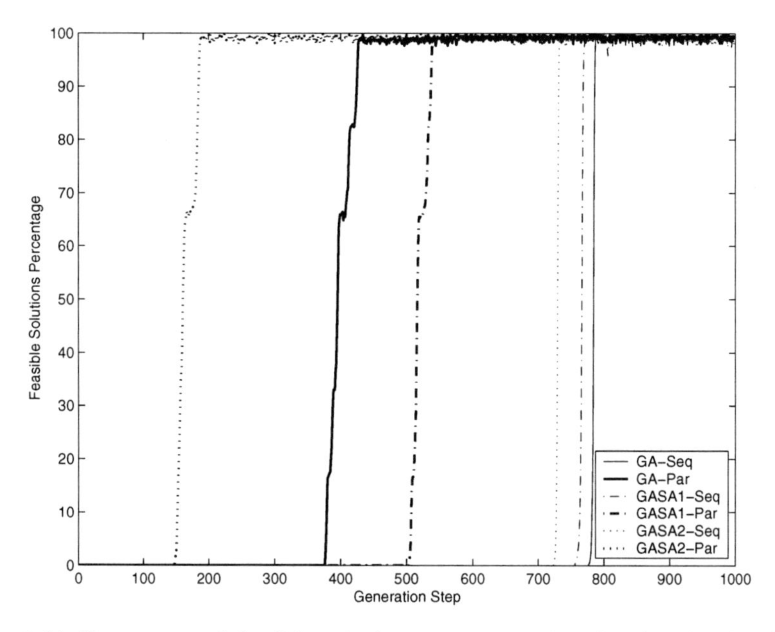


**Fig. 6.11** Percentage of feasible solutions per generation for the circuit of the fourth case study.

the best solution reported in the literature for this circuit [171], which has 7 gates and a fitness value of 114. The Boolean expression of best solution found for this case study is: $S_0 = E' + DC$, $S_1 = A' + BC$, $S_2 = C \oplus BC$. Note that GASA2 has a slightly better average performance than GASA1. The use of parallelism for this instance produced only a slight increase in the average fitness of GASA1 and GASA2 and also helped these two algorithms to increase their hit rate. In fact, the use of parallelism increased the average fitness of all the approaches compared. It is also worth noticing that the use of parallelism helped the GA to converge to the best-known solution for this circuit, although its hit rate was low (5%).

Fig. 6.12 shows the (average) percentage of feasible solutions present in the population over time (i.e., generations) for each of the algorithms compared.

**Table 6.9** Comparison of the best solutions found for the fourth case study by GASA2, the $n$-cardinality genetic algorithm (**NGA**) [147], a human designer (**HD 1**) who used Karnaugh maps and theorems from Boolean algebra, a second human designer (**HD 2**) who used the Quine-McCluskey method and **Miller et al.** [150], who used *cartesian genetic programming.*

| GASA2 | NGA | HD 1 | HD 2 | Miller et al. |
|---|---|---|---|---|
| 7 gates | 9 gates | 8 gates | 12 gates | 9 gates |

**Table 6.10** Comparison of results for the fifth case study.

| Algorithm | sequential | | | | parallel | | | |
|---|---|---|---|---|---|---|---|---|
| | opt | hits | avg | #evals | opt | hits | avg | #evals |
| GA | 113 | 5% | 100.20 | 933120 | **114** | 5% | 102.55 | 825603 |
| CHC | 102 | 5% | 89.35 | 546240 | 104 | 10% | 90.76 | 540632 |
| SA | 111 | 10% | 94.85 | 280883 | 112 | 5% | 98.64 | 256234 |
| GASA1 | **114** | 10% | 101.94 | 1013040 | **114** | 20% | 104.52 | 1010962 |
| GASA2 | **114** | 20% | 106.75 | 1382540 | **114** | **35%** | 106.90 | 1313568 |

In this case, the parallel version of the GA was the best performer (reaching 100% feasibility before generation 200), closely followed by both the sequential and the parallel versions of GASA1 (which reached 100% feasibility before generation 300). The worst performer was the sequential version of the GA. However, all the approaches were able to reach 100% feasibility. GASA2 was also the best overall performer in this case, reaching the highest average fitness. GASA2 also converged more consistently to the best-known solution.

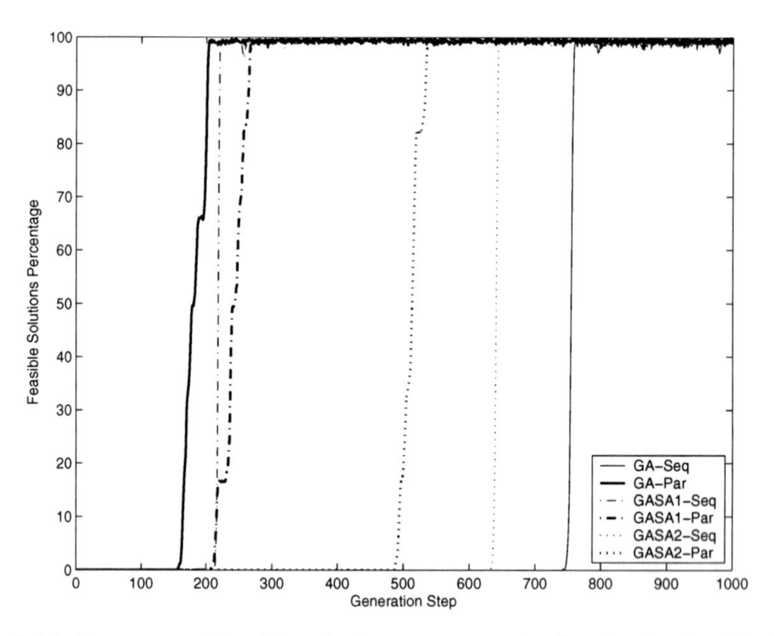


**Fig. 6.12** Percentage of feasible solutions per generation for the circuit of the fifth case study.

We show in Table 6.11 a comparison of the best solution found by GASA2 with respect to other approaches previously used to design the circuit of the fourth case study. This second comparison is only in terms of the Boolean expression found. In this case, GASA2 matched the solutions produced by a multiobjective genetic algorithm [173] and an approach based on particle swarm optimization [171]. Note that all of these approaches performed about one million fitness function evaluations each.

**Table 6.11** Comparison of the best solutions found for the fifth case study by GASA2, the multiobjective genetic algorithm (**MGA**) [173], and particle swarm optimization [171] (**PSO**).

| **GASA2** | **MGA** | **PSO** |
|---|---|---|
| 7 gates | 7 gates | 7 gates |

## 6.6 Overall Discussion

After this study, a few general conclusions can be inferred from our results. First, the hybridization of a genetic algorithm with simulated annealing seems to be beneficial for designing combinational logic circuits, at least when compared to pure separated GA and SA algorithms. From the two hybrids considered, GASA2 had the best performance. This is apparently due to its use of simulated annealing over the final population of GASA1, which allows to focus the search on more specific regions (something hard to do with the traditional genetic operators).

On the other hand, despite our belief that the highly disruptive recombination operator of CHC would be beneficial in circuit design, our results indicate that this approach has the worst overall performance of all the heuristics tried. Apparently, the mating restrictions of CHC (*incest prevention*) and its *restart* process were not sufficient to compensate for the lack of diversity due to its elitist selection, and the approach had difficulties to converge to feasible solutions.

SA also presented poor results compared to the hybrids and the GA. Although, in several problems, SA obtained similar final best fitness values as to the GA, its average fitness is often lower than the other methods. The reason for this is that SA rapidly finds a local optimum from which it can not escape, in spite of the internal mechanism explicitly added to the algorithm to avoid them. However, this method gets fairly accurate results with a fewer number of evaluations than the other algorithms.

Finally, we also found that, in most cases, the use of parallelism improves the average fitness of the approaches compared. This is something interesting, since it constitutes an additional motivation to parallelize the heuristics adopted to design combinational logic circuits. However, it was also found that this increase in the average fitness of the approaches was normally accompanied by a decrease in the hit rate. In other words, some consistency (or robustness) was sacrificed at the expense of achieving solutions of a higher quality.

## 6.7 Conclusions and Future Work

The comparative study conducted in this chapter has shown that the hybridization of an evolutionary algorithm with simulated annealing may bring benefits when designing combinational logic circuits. Emphasis is placed on the fact that the GA hybridized is using binary encoding. Additionally, the use of parallelism also brought benefits in terms of the quality of solutions produced, but it did not necessarily improve the hit rate (i.e., the number of times that an algorithm converged to its best found solution). Note also that the use of parallelism tended to decrease the average number of evaluations required by each algorithm to achieve their best possible fitness value. Nevertheless, a more in-depth study of the impact of parallelism in combinational circuit design remains as an open research area.

As part of potential future works, for example, it is interesting to use a population-based multiobjective optimization approach (the so-called MGA proposed in [146]) hybridized with a SA. Intuitively, this sort of approach should produce better results when hybridized, since by itself is a very powerful search engine for combinational circuit design. However, this approach will be quite time-consuming, and then, the utilization of parallel platforms will be needed to obtain results in a reasonable time.

# 7

# Parallel Genetic Algorithm for the Workforce Planning Problem

*I work quickly to live calmly.*

*Monserrat Caballé (1933 - ) - Spanish opera singer*

Decision making associated with workforce planning results in difficult optimization problems, this is so it involves multiple levels of complexity. In fact, the workforce planning problem that we tackle in this chapter consists of two sets of decisions: selection and assignment. The first step selects a small set of employees from a large number of available workers and the second (decision) assigns this staff to the tasks to be performed. The objective is to minimize the costs associated to the human resources needed to fulfill the work requirements. An effective workforce plan is an essential tool to identify appropriate workload staffing levels and justify budget allocations so that organizations can meet their objectives.

The complexity of this problem does not allow the utilization of exact methods for instances of realistic size. As a consequence, in this chapter we firstly propose a parallel genetic algorithm (GA) and we compare its performance with respect to another parallel metaheuristic: a parallel scatter search (SS). Two kinds of instances have been used to test our approaches. In "structured" ones, there exists a relationship between the tasks duration and the time that a worker can be assigned to them. On the other hand, this constraint is not considered in "unstructured" ones any more, turning these instances more difficult to solve. The development of these methods has the goal of providing a tool for finding high-quality solutions to structured and unstructured instances of the workforce planning problem (WPP). A preliminary set of results with these two algorithms, where the scatter search approaches outperformed the genetic algorithms [174, 175], have led us to begin to work on the hypothesis that the improvement operator of SS could be the key component provoking these enhancements. Therefore, in this chapter we present a new hybrid genetic algorithm in which this operator is applied (with certain probability) in its operator pool. The results will confirm our guess and represent a new state of the art tool for this benchmark.

The organization of this chapter is as follows. In next section we show a mathematical description of the WPP. In Section 7.2 and Section 7.3 we

G. Luque and E. Alba: Parallel Genetic Algorithms, SCI 367, pp. 115–134.
springerlink.com 

describe the parallel genetic algorithm and the parallel scatter search, respectively. Then, in Section 7.4 we analyze the results of these algorithms for the solution of the WPP, and finally, we give some hints on future works and conclusions in Section 7.5.

## 7.1 The Workforce Planning Problem

The following description of the problem is taken from Glover *et al.* [176]. A set of jobs $J = \{1, \ldots, m\}$ must be completed during the next planning period (e.g., a week). Each job $j$ requires $d_j$ hours during the planning period. There is a set $I = \{1, \ldots, n\}$ of available workers. The availability of worker $i$ during the planning period is $s_i$ hours. For reasons of efficiency, a worker must perform a minimum number of hours ($h_{min}$) of any job to which he/she is assigned and, at the same time, no worker may be assigned to more than $j_{max}$ jobs during the planning period. Workers have different skills, so $A_i$ is the set of jobs that worker $i$ is qualified to perform. No more than $t$ workers may be assigned during the planning period. In other words, at most $t$ workers may be chosen from the set $I$ of $n$ workers and the subset of selected workers must be capable of completing all the jobs. The goal is to find a feasible solution that optimizes a given objective function.

We use the cost $c_{ij}$ of assigning worker $i$ to job $j$ to formulate the optimization problem associated with this workforce planning situation as a mixed-integer program. We refer to this model of the workforce planning problem as WPP:

$$x_{ij} = \begin{cases} 1 \text{ if worker } i \text{ is assigned to job } j \\ 0 \text{ otherwise} \end{cases}$$

$$y_i = \begin{cases} 1 \text{ if worker } i \text{ is selected} \\ 0 \text{ otherwise} \end{cases}$$

$$\begin{aligned} z_{ij} &= \text{number of hours that worker } i \text{ is assigned to} \\ &\quad \text{perform job } j \\ Q_j &= \text{set of workers qualified to perform job } j \end{aligned}$$

$$\text{Minimize} \sum_{i \in I} \sum_{j \in A_i} c_{ij} \cdot x_{ij} \tag{7.1}$$

Subject to

$$\sum_{j \in A_i} z_{ij} \leq s_i \cdot y_i \qquad \forall i \in I \tag{7.2}$$

$$\sum_{i \in Q_j} z_{ij} \geq d_j \qquad \forall j \in J \tag{7.3}$$

$$\sum_{j \in A_i} x_{ij} \le j_{max} \cdot y_j \qquad \forall i \in I \tag{7.4}$$

$$h_{min} \cdot x_{ij} \le z_{ij} \le s_i \cdot x_{ij} \ \forall i \in I, j \in A_i \tag{7.5}$$

$$\sum_{i \in I} y_i \le t \tag{7.6}$$

$$x_{ij} \in \{0,1\} \ \forall i \in I, j \in A_i$$

$$y_i \in \{0,1\} \qquad \forall i \in I$$

$$z_{ij} \ge 0 \ \forall i \in I, j \in A_i$$

In the model above, the objective function (Equation 7.1) minimizes the total assignment cost. Constraint set (Equation 7.2) limits the number of hours for each selected worker. If the worker is not chosen, then this constraint does not allow any assignment of hours to him/her. Constraint set (Equation 7.3) enforces the job requirements, as specified by the number of hours needed to complete each job during the planning period. Constraint set (Equation 7.4) limits the number of jobs that a chosen worker is allowed to perform. Constraint set (Equation 7.5) enforces that once a worker has been assigned to a given job, he/she must perform such a job for a minimum number of hours. Also, constraint (Equation 7.5) does not allow the assignment of hours to a worker that has not been chosen to perform a given job. Finally, constraint set (Equation 7.6) limits the number of workers chosen during the current planning period.

The same model may be used to optimize a different objective function. Let $\hat{c}_{ij}$ be the cost per hour of worker $i$ when performing job $j$. Then, the following objective function minimizes the total assignment cost (on hourly basis):

$$\text{Minimize} \sum_{i \in I} \sum_{j \in A_i} \hat{c}_{ij} \cdot z_{ij} \tag{7.7}$$

Alternatively, $p_{ij}$ may reflect the preference of worker $i$ for job $j$, and therefore, the following objective function maximizes the total preference of the assignment:

$$\text{Maximize} \sum_{i \in I} \sum_{j \in A_i} p_{ij} \cdot x_{ij} \tag{7.8}$$

When preference values are used, other objective functions may be formulated. For instance, it may be desirable to maximize the minimum preference value for the set of selected workers. In this chapter, we assume that the decision maker wants to minimize the total assignment costs as calculated in Equation 7.1.

As pointed out in [177], this problem is related to the capacitated facility location problem (CFLP) as well as the capacitated $p$-median problem [178, 179]. In fact, our location-allocation problem reduces to a CFLP if the complicating constraints (Equations 7.4-7.6) are relaxed in a Lagrangean manner. In the context of the CFLP, implied bounds are typically added to strengthen the linear programming (LP) relaxation of the mixed-integer programming formulation. The equivalent bounds for the WPP formulation are:

$$x_{ij} - y_i \leq 0 \qquad \forall i \in I, j \in A_i \tag{7.9}$$

Also in the case of the CFLP, an aggregate capacity constraint is usually added to the problem formulation in order to improve some Lagrangean bounds. Even in the case of an LP approach this surrogate constraint can be helpful; it can be used for generating possibly violated lifted cover inequalities. The form of such a constraint for the WPP model is:

$$\sum_{i \in I} s_i \cdot y_i \geq \sum_{j \in J} d_j \tag{7.10}$$

The difficulty of solving instances of the WPP with an optimization method is related to the relationship between $h_{min}$ and $d_j$. In particular, problem instances for which $d_j$ is a multiple of $h_{min}$ (referred to as "structured") are easier to handle than those for which $d_j$ and $h_{min}$ are unrelated (referred to as "unstructured").

## 7.2 Design of a Genetic Algorithm

A genetic algorithm (GA) [162] is an iterative technique that applies stochastic operators on a pool of individuals (the population). Every individual in the population is the encoded version of a tentative solution. Initially, this population is randomly generated. An evaluation function associates a fitness value to every individual indicating its suitability to the problem.

The genetic algorithm that we have developed for the solution of the WPP follows the basic structured shown in Fig. 7.1. The population size for GAs used in this work is 400 individuals. This value and the specific values of the parameters in the following sections as well have been obtained after preliminary experimentation.

Within the basic structure of the GA for solving the WPP, we have added context information through a special solution representation and crossover operators with improving and repairing mechanisms.

### *7.2.1 Solution Encoding*

Solutions are represented as an $n \times m$ matrix $Z$, where $z_{ij}$ represents the number of hours that worker $i$ is assigned to job $j$. In this representation, a worker $i$ is considered to be assigned to job $j$ if $z_{ij} > 0$. A solution using this representation is showed in Fig. 7.2. Therefore the following relationships are established from the values in $Z$.

```
Generate (P(0))
t := 0
while not Termination_Criterion(P(t)) do
    Evaluate(P(t))
    P'(t) := Selection(P(t))
    P'(t) := Recombination(P'(t))
    P'(t) := Mutation(P'(t))
    P(t+1) := Replacement(P(t), P'(t))
    t := t+1
return Best_Solution_Found
```

**Fig. 7.1** Basic GA structure.

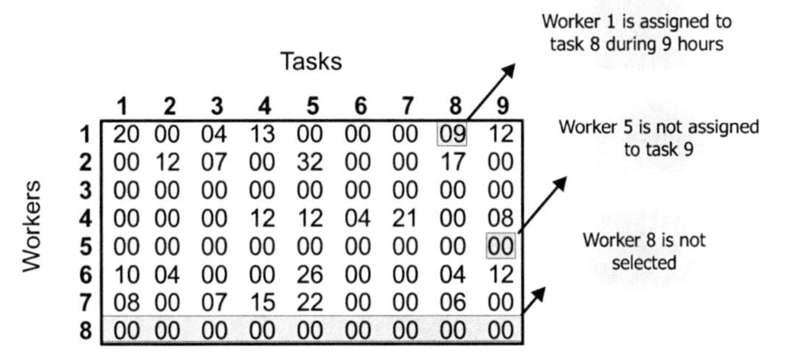


**Fig. 7.2** An example of a solution using the representation presented in this work.

$$x_{ij} = \begin{cases} 1 \text{ if } z_{ij} > 0 \\ 0 \text{ otherwise} \end{cases}$$

$$y_i = \begin{cases} 1 \text{ if } \sum_{j \in A_i} z_{ij} > 0 \\ 0 \text{ otherwise} \end{cases}$$

### *7.2.2 Evaluation the Quality of a Solution*

Solutions are evaluated according to the objective function (7.1) plus a penalty term. The additional term penalizes violations of constraints (equations 7.2, 7.3, 7.4 and 7.6). The penalty coefficients that are multiplied by the constraint violations are $p_2$, $p_3$, $p_4$, and $p_6$. Values for these coefficients have been set up to 50, 50, 200, and 800, respectively. Before the fitness value is calculated, new trial solutions undergo a repairing/improving operation that makes sure that constraint of Equation 7.5 is satisfied. This operator also ensures that no worker is assigned to a job that he/she is not qualified to perform.

### *7.2.3 Repairing/Improving Operator*

The purpose of this operator is to repair trial solutions in such a way that they either become feasible with respect to the original problem or at least the infeasibility of these solutions is reduced. The operator performs the 4 steps outlined in Fig. 7.3.

In the first step, this operator repairs solutions with respect to the minimum number of hours that a worker must work on any assigned job. The repair is done only on those qualified workers that are not meeting the minimum time requirement. In mathematical terms, if $0 < z_{ij} < h_{min}$ for $i \in I$ and $j \in A_i$, then $z_{ij} = h_{min}$.

The second step takes care of assignments of workers to jobs for which they are not qualified to perform. A value of zero is given to the corresponding entry in $Z$. Using our notation, if $z_{ij} > 0$ for $i \in I$ and $j \notin A_i$, then $z_{ij} = 0$.

The third step considers that a worker is feasible if he/she satisfies constraints of equations 7.2 and 7.4. This step attempts to use up the capacity slack of feasible workers. The slack time for worker $i$ (i.e., $s_i - \sum_{j \in A_i} z_{ij}$ ) is equally divided among his/her current job assignments. This allows for a higher utilization of the workers that are currently assigned to jobs and thus facilitating the satisfaction of constraint of equation 7.6.

| |
|---|
| 1. Eliminate violations with respect to $h_{min}$ |
| 2. Eliminate violations with respect to assignments of unqualified workers |
| 3. Load feasible workers |
| 4. Reduce infeasibility |

**Fig. 7.3** Repairing/improving operator.

The last step starts with a partial order of the workers in such a way that those which provoke the largest constraint violation with respect to constraints (7.2) and (7.4) tend to appear at the top of the list. This is not a complete order relationship because the operator accounts for a certain amount of randomness in this step. Once the partial order is established, a process of reducing the infeasibility of workers is applied. The process of reducing the violation of constraints (7.2) and (7.4) is only applied if it does not provoke new violations of constraints (3) and (5).

### *7.2.4 Recombination Operator*

A special recombination operator has been designed for the solution of WPP. The operator employs a parameter $\rho_c$ that may be interpreted as the probability that two solutions exchange their current assignments for worker $i$. The process is summarized in Fig. 7.5 and an example is showed in Fig. 7.4.

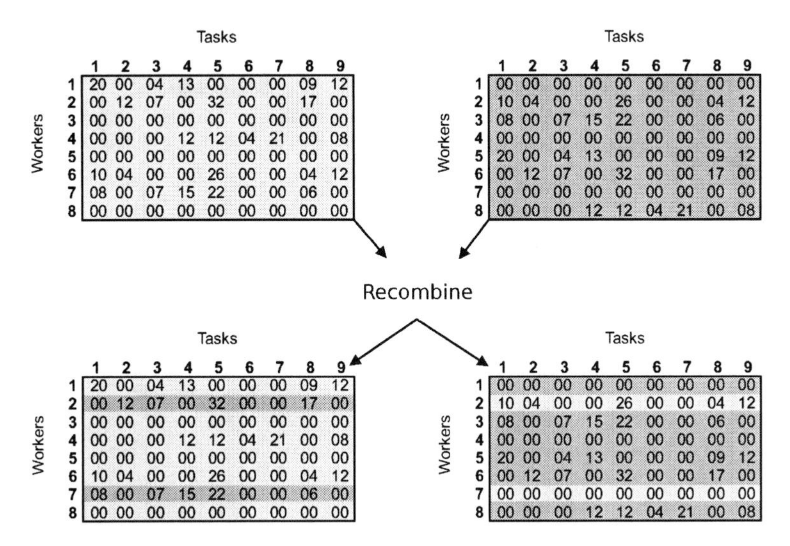


**Fig. 7.4** An example of application of the crossover operator.

```
for (i = 1 to n) do
    if rand() < ρc then
        for (j = 1 to m) do
            z1_ij ↔ z2_ij
        endfor
    endif
endfor
```

**Fig. 7.5** Crossover operator.

Given two solutions $Z1$ and $Z2$, the recombination operator in Fig. 7.5 selects, with probability $\rho_c$, a worker $i$. In the experimentation section, this value is set up to 0.8. If the worker is selected, then the job assignments of solution $Z1$ are exchanged with the assignments of solution $Z2$. The *rand()* function in Fig. 7.5 generates a uniform random number between 0 and 1.

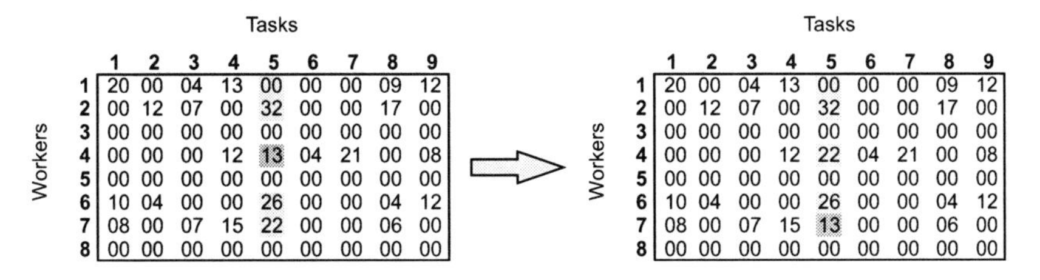


**Fig. 7.6** An example of application of the mutation operator. In each step of this operator, the current task assignment (dark-gray color) is randomly exchanged with the task assignment of other qualified worker (light-gray color).

### *7.2.5 Mutation Operator*

In addition to the crossover operator described before, our GA implementation includes a mutation operator. This mechanism operates on a single solution by exchanging the job assignments of two workers. The job exchange occurs with probability $\rho_m$, as shown in Fig. 7.7. An example of this operator is showed in Fig. 7.6.

Given a solution $Z$, the mutation operator considers all workers and jobs that the workers are qualified to perform. A random worker $k$ is chosen from the list of qualified workers and the exchange of job assignments is performed. For the experiments, we set up $\rho_m$ to 0.2. As before, the $rand()$ function returns a uniform random number between 0 and 1.

### *7.2.6 The Proposed Parallel GA*

A parallel GA (pGA) [6] is a procedure that consists of multiple copies of an implementation (typically serial) of a genetic algorithm. The individual GAs include an additional communication phase that enables them to exchange information. A pGA is characterized by the nature of the individual GAs and the type of communication that is established among them. Our particular implementation is a distributed GA (dGA), which allows for an efficient exploitation of machine clusters. Typically, dGAs consist of a small number of independent GAs that periodically exchange information. Each individual GA operates on a considerably large population. Since we want to compare against the sequential GA, pGAs use the same population size, but now the whole population of the sequential GA is split into as many subpopulations as processes involved in the parallel computation.

To fully characterize a dGA, the migration policy must be established, which is related to the connection topology of the set of individual GAs. The policy dictates when migration occurs, the number and identity of the individuals that will be exchanged and also determines the synchronization scheme. Our implementation uses a unidirectional ring topology, where each GA receives information from the GA immediately preceding it and sends

**for** $(i = 1$ **to** $n)$ **do**
    **for** $(j \in A_i)$ **do**
        $k =$ random worker $|k \neq i$ **and** $k \in Q_j$
        **if** $rand() < \rho_m$ **then**
            $z_{ij} \leftrightarrow z_{kj}$
        **endif**
    **endfor**
**endfor**

**Fig. 7.7** Mutation operator.

```
generate P
build RefSet from P
while not end do
    generate subsets S from RefSet
    for each subset s in S do
        recombine solution in s to obtain x_s
        improve x_s
        Update RefSet with x_s
    if convergence(RefSet)
        generate a new P
        build Refset from P and the old RefSet
```

**Fig. 7.8** Basic SS structure.

information to the GA that is immediately after it. At each migration operation (which is carried out every 15 generations), one single solution is selected from the population (via binary tournament) and sent to the corresponding neighbor. The newly reached solution replaces the worst individual in the target population only if it is better.

## 7.3 Scatter Search

Scatter Search (SS) [180] is also a population-based metaheuristic that uses a reference set to combine its solutions and construct others. The method generates a reference set from a population of solutions. Then a subset is selected from this reference set. The selected solutions are combined to get starting solutions to run an improvement procedure. The result of this improvement can motivate the updating of the reference set. A pseudo-code of the scatter search is showed in Fig. 7.8. The procedures involved by the SS method are the following:

- *Initial population creation:* The first step of this technique is to generate an initial population which is the base set to build the Reference Set. This population must be a wide set of disperse (non similar) solutions. However, it must clearly also include good quality solutions. A simple method to create this population is to use a random generation one (disperse solutions) and then improving some of the solutions to obtain high quality ones. However, other several strategies could be applied to get a population with these properties using problem (heuristic) information.
- *Reference Set update and creation:* The SS operates on a small set of solutions, the *RefSet*, consisting of the "good" solutions found during the search. The "good" solutions are not limited to those with the best objective values. By "good" solutions we mean solutions with the best objective values as well as disperse solutions (to escape local optimality and diversify the search). In general, the RefSet is composed of two separate subsets: one

subset for the best solutions ($\text{RefSet}_1$) and another for diverse solutions ($\text{RefSet}_2$). This reference set is created from the initial population and updated when a new solution is generated. Also, this set is partially reinitialized when the search has stagnated. In our experiments, we use an small RefSet composed of eight solutions ($|RefSet_1| = 5$ and $|RefSet_2| = 3$).
- *Subset generation:* This procedure operates in the reference set to produce a subset of its solutions as a basis for creating combined solutions. In this work, we generate all 2-elements subsets (28 subsets) and then we apply the solution combination operator to them.
- *Solution combination:* It transforms a given subset of solutions into one or more combined solution vectors.
- *Improvement method:* This procedure transforms the current solution into an enhanced solution.

To solve the WPP with SS, we have used the same representation, fitness evaluation, repairing operator, and crossover operator that we used with GA implementation (Sections 7.2.1, 7.2.2, 7.2.3, and 7.2.4, respectively). These operators have been utilized because they perform an exhaustive and structured search but with new ideas extending SS. The rest of implementation issues are described in the next subsections.

### *7.3.1 Seeding the Initial Population*

In our case, the initial population is composed of 15 random solutions which are later enhanced by the improvement method described in the next subsection and then inserted into the initial population. As in the GA, we want to remark here that the entire parameterization of SS has been tuned properly after preliminary experimentation.

### *7.3.2 Improvement Method*

A special improvement operator has been designed for the solution of the WPP. The operator employs a parameter $\rho_i$ that may be interpreted as the probability that a worse solution replaces a better solution in the improvement method. The process is summarized in Fig. 7.9.

```
for (i = 1 to MaxIter) do
    Z' = generate neighbor from Z
    if fitness(Z') < fitness(Z) or
       rand() < ρ_i then
        Z = Z'
    endif
endfor
```

**Fig. 7.9** Improvement operator.

Given a solution $Z$, the improvement operator generates a neighbor (we use the mutation operator described in Section 7.2.5). If this new solution $Z'$ is better than the original solution $Z$, we accept that solution and the process is repeated for $MaxIter$ iterations. This method also accepts a worse solution by means of a probability defined by $\rho_i$. As before, the $rand()$ function returns a uniform random number between 0 and 1, and $fitness()$ returns the objective fitness value achieved by a solution. Our SS algorithm will perform 50 iterations of this process and the probability of accepting a worse solution ($\rho_i$) is 0.1 to escape from local optima.

### 7.3.3 Parallel SS

Several parallel implementations of the basic scheme of SS have been proposed in the literature [181]. We are interested in obtaining a parallel method which allows us not only to reduce the execution time but also improve the solution quality. Hence, we rule out the master-slave model since its numerical performance is the same as the sequential one.

We have used a distributed model, i.e., we have several sequential SS running in parallel that periodically exchange information (one single solution from RefSet). The connection topology is the same as in the pGA. Binary tournament is used for choosing the migrant, which promotes high quality solutions from $\text{RefSet}_1$ likely to be selected. In the target SS algorithm, the method for updating the RefSet is applied in order to insert the migrant solution.

In this method, the number of evaluation performed in each step is related to the number of subsets generated. Therefore, we reduce the number of subsets generated by each independent SS so that the computational effort is the same as the sequential version. In concrete, the number of subsets generated is the number of subsets of the serial version (a predefined value) divided by the number of islands. In this case, we choose the subset randomly, but we do not allow the same subset to be selected two or more times.

## 7.4 Computational Experiments and Analysis of Results

In this section we first present the problem instances used. Then, we analyze the behavior of the algorithms with respect to, on the one hand, their ability to find accurate solutions and, on the other hand, the time needed to reach these solutions.

The algorithms in this work have been implemented in C++ and executed on a cluster of Pentium 4 at 2.8 GHz with 512 MB of memory which run SuSE Linux 8.1 (kernel 2.4.19-4GB). The interconnection network is a Fast-Ethernet at 100 Mbps.

### 7.4.1 Problem Instances

In order to test the merit of the proposed procedure, we generated artificial problem instances. Given the values of $n$, $m$, and $t$ the problem instances were generated with the following characteristics:

$$\begin{aligned}
s_i &= U(50, 70) \\
j_{max} &= U(3, 5) \\
h_{min} &= U(10, 15) \\
\text{Category(worker i)} &= U(0, 2) \\
P(i \in Q_j) &= 0.25 \cdot (1 + \text{Category(worker i)}) \\
d_j &= max\left(h_{min}, U\left(\tfrac{\bar{s}\cdot t}{2\cdot m}, \tfrac{1.5\cdot\bar{s}\cdot t}{m}\right)\right) \\
\text{where } \bar{s} &= \tfrac{\sum_i s_i}{n} \text{ and } \tfrac{\sum_j d_j}{\bar{s}\cdot t} \leq \alpha \\
c_{ij} &= |A_i| + d_j + U(10, 20)
\end{aligned}$$

The generator establishes a relationship between the flexibility of a worker and his/her corresponding cost (salary). That is, workers that are able to perform more jobs are more expensive. We solve 10 structured plus other 10 unstructured problems which have been called $s1$ to $s10$ and $u1$ to $u10$, respectively. The ten unstructured problems were generated with the following parameter values: $n = 20$, $m = 20$, $t = 10$ and $\alpha = 0.97$.

Note that the problem generator uses $\alpha$ as the limit for the expected relative load of each worker. The set of ten structured problems was constructed using the same parameter values but $h_{min}$ was set to 4 and the $d_j$ values were adjusted as follows: $d_j = d_j - mod(d_j, 4)$, where $mod(x, y)$ calculates the remainder of $x/y$. All twenty instances were generated in such a way that a single value for the total number of available hours ($s_i$) is drawn and assigned to all workers.

### 7.4.2 Results: Workforce Planning Performance

The resulting workforce plannings computed by both GA and SS approaches are analyzed in this section. All the algorithms stop after 800,000 function evaluations have been computed. This guarantees that they all are able to converge. The configuration setting is showed in Table 7.1. These values were obtained experimentally. Values in the tables are average results over 30 independent runs. Since we deal with stochastic algorithms, we have carried out an statistical analysis of the results which consists of the following steps. First a Kolmogorov-Smirnov test is performed in order to check whether the variables are normal or not. Since all the Kolmogorov-Smirnov normality tests in this work were not successful, we use a non-parametric test: Kruskal-Wallis (with 95% of confidence).

We here remark that the parallel versions of GA and SS have been executed not only in parallel, but also on a single processor. The first reason that motivates these experiments is to check that the parallel search model is

**Table 7.1** Parameter settings.

| parameter | GA | SS |
|---|---|---|
| **Population size** | 400 | Initial Pop. = 15<br>RefSet = 8 |
| $\rho_c$ | 0.8 | - |
| $\rho_m$ | 0.2 | - |
| **Subset generated** | - | all 2-elements subsets (28) |
| $\rho_i$ | - | 0.1 |

**Table 7.2** GA results for structured and unstructured problems.

| Prob. | Seq. GA | pGA-4 | | | pGA-8 | | | $KW_3$ |
|---|---|---|---|---|---|---|---|---|
| | | 1 p. | 4 p. | $KW_2$ | 1 p. | 8 p. | $KW_2$ | |
| s1 | 963 | 880 | 879 | – | **873** | **873** | – | + |
| s2 | 994 | 943 | 940 | – | **920** | 922 | – | + |
| s3 | 1156 | **1013** | 1015 | – | 1018 | 1016 | – | + |
| s4 | 1201 | 1036 | 1029 | – | 1008 | **1003** | – | + |
| s5 | 1098 | 1010 | 1012 | – | **998** | 1001 | – | + |
| s6 | 1193 | 1068 | 1062 | – | **1042** | 1045 | – | + |
| s7 | 1086 | 954 | 961 | – | 960 | **953** | – | + |
| s8 | 1287 | 1095 | 1087 | – | **1068** | 1069 | – | + |
| s9 | 1107 | **951** | 956 | – | 984 | 979 | – | + |
| s10 | 1086 | 932 | 927 | – | **924** | 926 | – | + |
| u1 | 1631 | 1386 | 1372 | – | **1302** | 1310 | – | + |
| u2 | 1264 | 1132 | **1128** | – | 1153 | 1146 | – | + |
| u3 | 1539 | **1187** | 1193 | – | 1254 | 1261 | – | + |
| u4 | 1603 | 1341 | 1346 | – | 1298 | **1286** | – | + |
| u5 | 1356 | **1241** | 1252 | – | 1254 | 1246 | – | + |
| u6 | 1205 | 1207 | 1197 | – | 1123 | **1116** | – | + |
| u7 | 1301 | 1176 | 1179 | – | 1127 | **1121** | – | + |
| u8 | **1106** | 1154 | 1151 | – | 1123 | 1128 | – | – |
| u9 | 1173 | 950 | 938 | – | **933** | 935 | – | + |
| u10 | 1214 | **1160** | 1172 | – | 1167 | 1163 | – | + |

independent of the computing platform. As expected, the corresponding tests included in the $KW_2$ columns of Tables 7.2 and 7.3 indicate that no statistical difference exists between them ("–" symbols). In effect, this confirms that they are the same numerical model. As a consequence, in order to compare sequential *vs.* parallel versions of each algorithm, we have considered only the results of the parallel executions of pGA and pSS (i.e., we do not use the 1p. columns) and therefore the statistical test just involves three datasets (column $KW_3$). In the second place, running parallel models in a single CPU will allow us to perform the execution time analysis of the algorithms properly (see Section 7.4.3 for the details). The result of the best algorithm for each instance is marked in **boldface**. Let us discuss them in separate sections.

### GA Results

The first conclusion that can be drawn from Table 7.2 is that any pGA configuration is able to solve the considered WPP better than the sequential GA, and statistical confidence exists for this claim (see "+" symbols in column

**Table 7.3** SS results for structured and unstructured problems.

| Prob. | Seq. SS | pSS-4 | | | pSS-8 | | | $KW_3$ |
|---|---|---|---|---|---|---|---|---|
| | | 1 p. | 4 p. | $KW_2$ | 1 p. | 8 p. | $KW_2$ | |
| s1 | 939 | 896 | 901 | – | **861** | 862 | – | + |
| s2 | 952 | **904** | 905 | – | 916 | 913 | – | + |
| s3 | 1095 | 1021 | 1019 | – | 1005 | **1001** | – | + |
| s4 | 1043 | 1002 | **991** | – | 997 | 994 | – | + |
| s5 | 1099 | **999** | 1007 | – | 1009 | 1015 | – | + |
| s6 | 1076 | 1031 | 1034 | – | 1023 | **1022** | – | + |
| s7 | 987 | 956 | 942 | – | 941 | **933** | – | + |
| s8 | 1293 | 1113 | 1120 | – | **1058** | 1062 | – | + |
| s9 | 1086 | **948** | 950 | – | 952 | 950 | – | + |
| s10 | 945 | **886** | 891 | – | 915 | 909 | – | + |
| u1 | 1586 | 1363 | 1357 | – | 1286 | **1280** | – | + |
| u2 | 1276 | 1156 | 1158 | – | 1083 | **1078** | – | + |
| u3 | 1502 | 1279 | 1283 | – | **1262** | 1267 | – | + |
| u4 | 1653 | 1363 | 1356 | – | 1307 | **1305** | – | + |
| u5 | 1287 | 1176 | 1192 | – | 1175 | **1169** | – | + |
| u6 | 1193 | 1168 | 1162 | – | 1141 | **1136** | – | – |
| u7 | 1328 | 1152 | 1151 | – | 1084 | **1076** | – | + |
| u8 | 1141 | 1047 | 1039 | – | **1031** | 1033 | – | + |
| u9 | 1055 | 906 | 908 | – | 886 | **883** | – | + |
| u10 | 1178 | 1003 | 998 | – | **952** | 958 | – | + |

$KW_3$). The unstructured problem u8 stands for the exception but it can be ruled out since the Kruskal-Wallis test is negative ("–" symbol in column $KW_3$), thus indicating that the algorithms are not statistically different from each other. Specially accurate solutions have been computed by the pGA in unstructured instances $u1$, $u3$, and $u9$, where the reductions in the planning costs are above 20%.

If we now compare pGAs among them, Table 7.2 shows that pGA-8 found the best solutions for 13 out of 20 WPP instances, while pGA-4 was only able to find the best plannings in 6 out of 20. This holds specially for the structured problems where pGA-8 gets the best workforce plannings in 8 out of 10 instances. However, it is also noticeable that differences between solutions from pGA-4 and pGA-8 are very small, thus showing that both algorithms have a similar ability for solving the WPP.

### SS Results

We can start analyzing the results of SS (Table 7.3) in the same way as GA results and state the same conclusions, i.e., parallel SS configurations always get the best solutions versus SS for all the WPP instances and also with statistical confidence ("+" symbols in column $KW_3$). There are some particular instances in which pSS was able to reduce the planning costs significantly with respect to the sequential SS, e.g. s8, from 1293 down to 1048 (reduction of 18%) or u4, from 1653 down to 1305 (reduction of 21%). Averaging over structured and unstructured instances, the best pSS configuration reduces WPP costs of sequential SS in 8.35% and 14.98%, respectively.

Turning to compare pSS-4 and pSS-8 between them, Table 7.3 shows that no conclusion can be draw concerning the structured problems since both algorithms get the best solutions for 5 out of 10 instances. However, pSS-8 always reaches the best workforce planning in the case of the unstructured problems, so (as with pGA-8) we can conclude a slight advantage of pSS-8 over pSS-4.

**GA vs. SS**

In this section we want to compare both GA and SS approaches for solving WPP. Since there are many different problem instances and analyzing them thoroughly would hinder us from drawing clear conclusions, we have summarized in Table 7.4 the information of Tables 7.2 and 7.3 as follows: we have normalized the resulting planning cost for each problem instance with respect to the worst (maximum) cost obtained by any proposed algorithm, so we can easily compare without scaling problems. Then, values in Table 7.4 are average values over all the structured and unstructured WPP instances.

A clear conclusion that can be reached is that all SS configurations outperform the corresponding GA ones, that is, considering all structured and unstructured WPP instances, SS gets better solutions than the GA. It is worth mentioning differences between sequential approaches in structured problems (normalized average is reduced from 0.9994 down to 0.9410) and eight island based parallel algorithm in unstructured problems, where pSS-8 normalized costs are 4.4% lower than pGA-8 ones. These results allow us to conclude that SS is a more promising approach for solving this workforce planning problem. Although it can be explained because of the search model of SS by itself, we want to thoroughly discuss this fact. We conjecture that the improvement operator of SS could be responsible for such enhancements since adjusting the number of iterations that it performs was the most sensitive parameter in the preliminary experimentation. We will investigate on this in Section 7.4.4. Now, let us continue with our analysis, this time from the wall-clock point of view.

### *7.4.3 Results: Computational Times*

In order to have a fair and meaningful values of the metrics when dealing with such stochastic algorithms, we need to consider exactly the same algorithm

**Table 7.4** Average results for structured and unstructured problems.

| Problems | | s1 - s10 | | u1 - u10 | |
|---|---|---|---|---|---|
| Algorithm | | GA | SS | GA | SS |
| Sequential | | 0.9994 | 0.9410 | 0.9896 | 0.9744 |
| 4 Islands | 1 p. | 0.8858 | 0.8743 | 0.8885 | 0.8605 |
| | 4 p. | 0.8847 | 0.8747 | 0.8879 | 0.8598 |
| 8 Islands | 1 p. | 0.8783 | 0.8677 | 0.8735 | 0.8308 |
| | 8 p. | 0.8776 | 0.8663 | 0.8718 | 0.8292 |

**Table 7.5** Execution time (in seconds) for structured and unstructured problems.

| | | | | 4 Islands | | | | | | 8 Islands | | | | | | |
|---|---|---|---|---|---|---|---|---|---|---|---|---|---|---|---|---|
| | Sequential | | | 1 CPU | | | 4 CPUs | | | 1 CPU | | | 8 CPUs | | | $KW_6$ |
| Pbm | GA | SS | $KW_2$ | GA-4 | SS-4 | $KW_2$ | GA-4 | SS-4 | $KW_2$ | GA-8 | SS-8 | $KW_2$ | GA-8 | SS-8 | $KW_2$ | |
| s1 | 61 | 72 | + | 62 | 74 | + | 17 | 19 | + | 66 | 77 | + | 9 | 10 | + | + |
| s2 | 32 | 49 | + | 32 | 53 | + | 9 | 14 | + | 37 | 58 | + | 6 | 8 | + | + |
| s3 | 111 | 114 | − | 113 | 118 | + | 29 | 31 | + | 115 | 127 | + | 15 | 17 | + | + |
| s4 | 87 | 86 | − | 93 | 84 | + | 24 | 23 | − | 95 | 87 | + | 13 | 13 | − | + |
| s5 | 40 | 43 | − | 41 | 45 | + | 13 | 12 | − | 46 | 47 | − | 9 | 7 | + | + |
| s6 | 110 | 121 | + | 109 | 122 | + | 34 | 33 | − | 114 | 128 | + | 18 | 18 | − | + |
| s7 | 49 | 52 | + | 53 | 47 | + | 16 | 14 | + | 57 | 55 | − | 9 | 8 | + | + |
| s8 | 42 | 46 | − | 45 | 48 | − | 13 | 13 | − | 48 | 50 | − | 7 | 7 | − | + |
| s9 | 67 | 70 | + | 73 | 71 | − | 21 | 19 | + | 76 | 74 | − | 13 | 10 | + | + |
| s10 | 102 | 105 | + | 105 | 101 | + | 28 | 28 | − | 109 | 106 | + | 16 | 15 | − | + |
| u1 | 95 | 102 | + | 98 | 108 | + | 29 | 29 | − | 102 | 111 | + | 16 | 16 | − | + |
| u2 | 87 | 94 | + | 89 | 95 | + | 28 | 26 | + | 92 | 99 | + | 15 | 14 | − | + |
| u3 | 51 | 58 | + | 55 | 55 | − | 17 | 17 | − | 59 | 59 | − | 10 | 11 | + | + |
| u4 | 79 | 83 | + | 79 | 86 | + | 26 | 24 | + | 86 | 92 | + | 15 | 15 | − | + |
| u5 | 57 | 62 | + | 62 | 62 | − | 21 | 18 | + | 63 | 68 | + | 12 | 10 | + | + |
| u6 | 75 | 111 | + | 72 | 115 | + | 20 | 30 | + | 70 | 119 | + | 13 | 16 | + | + |
| u7 | 79 | 80 | − | 81 | 81 | − | 24 | 24 | − | 89 | 83 | + | 15 | 14 | − | + |
| u8 | 89 | 123 | + | 88 | 118 | + | 23 | 35 | + | 92 | 123 | + | 14 | 20 | + | + |
| u9 | 72 | 75 | − | 78 | 77 | − | 22 | 22 | − | 85 | 80 | + | 13 | 12 | − | + |
| u10 | 95 | 99 | + | 96 | 96 | − | 25 | 28 | − | 99 | 101 | + | 13 | 17 | + | + |

and then only change the number of processors, because comparing against the sequential versions would lead to misleading results [6]. This way, we have also executed parallel versions of both GA and SS also in a single CPU as shown in Table 7.5, where we include the average execution times at which the best solution is found during the computation of all the algorithms over 30 independent runs. The same statistical tests have been performed as in the previous section.

If we analyze the execution times of those algorithms being run on a single CPU, it can be seen that sequential optimizers are faster than the monoprocessor execution of any of their parallel version. In order to provide this claim with confidence, we include in column $KW_6$ the result of the statistical test using all the results computed with one single CPU. The "+" symbols in this column indicate that all the execution times are different with statistical significance. This holds for 17 out of 20 instances and 15 out of 20 ones in GA and SS, respectively. The overload of running the several processes of the parallel versions on a single CPU is the main reason for their lower execution. However, sequential algorithms for instances $s6$ and $u8$ in GAs and $s4$, $s7$, $s10$, $u3$, and $u8$ in SS obtain longer execution times than the parallel versions with 4 islands. The point here is that a trade-off exists between the overload due to the number of processes and the ability of the algorithms to easily reach the optimal solution. While the former issue tends to increase the computational time, the latter is a way of reduce it. Results in both tables point out that the computing overload is a very important factor because sequential algorithms usually perform faster than parallel algorithms on one processor.

Analyzing the absolute execution times, one can see the GAs generally get lower execution times than SS algorithms when the computing platform is

**Table 7.6** Parallel efficiency and serial fraction for structured and unstructured problems.

| | pGA-4 | | pSS-4 | | pGA-8 | | pSS-8 | |
|---|---|---|---|---|---|---|---|---|
| Problem | $\eta$ | $sf$ | $\eta$ | $sf$ | $\eta$ | $sf$ | $\eta$ | $sf$ |
| s1 | 0.91 | 0.032 | 0.97 | 0.009 | 0.91 | 0.012 | 0.96 | 0.005 |
| s2 | 0.88 | 0.041 | 0.94 | 0.018 | 0.77 | 0.042 | 0.90 | 0.014 |
| s3 | 0.97 | 0.008 | 0.95 | 0.016 | 0.95 | 0.006 | 0.93 | 0.010 |
| s4 | 0.96 | 0.010 | 0.91 | 0.031 | 0.91 | 0.013 | 0.83 | 0.027 |
| s5 | 0.78 | 0.089 | 0.93 | 0.022 | 0.63 | 0.080 | 0.83 | 0.027 |
| s6 | 0.80 | 0.082 | 0.92 | 0.027 | 0.79 | 0.037 | 0.88 | 0.017 |
| s7 | 0.82 | 0.069 | 0.83 | 0.063 | 0.79 | 0.037 | 0.85 | 0.023 |
| s8 | 0.86 | 0.051 | 0.92 | 0.027 | 0.85 | 0.023 | 0.89 | 0.017 |
| s9 | 0.86 | 0.050 | 0.93 | 0.023 | 0.73 | 0.052 | 0.92 | 0.011 |
| s10 | 0.93 | 0.022 | 0.90 | 0.036 | 0.85 | 0.024 | 0.88 | 0.018 |
| u1 | 0.84 | 0.061 | 0.93 | 0.024 | 0.79 | 0.036 | 0.86 | 0.021 |
| u2 | 0.79 | 0.086 | 0.91 | 0.031 | 0.76 | 0.043 | 0.88 | 0.018 |
| u3 | 0.80 | 0.078 | 0.80 | 0.078 | 0.73 | 0.050 | 0.67 | 0.070 |
| u4 | 0.75 | 0.105 | 0.89 | 0.038 | 0.71 | 0.056 | 0.76 | 0.043 |
| u5 | 0.73 | 0.118 | 0.86 | 0.053 | 0.65 | 0.074 | 0.85 | 0.025 |
| u6 | 0.90 | 0.037 | 0.95 | 0.014 | 0.67 | 0.069 | 0.92 | 0.010 |
| u7 | 0.84 | 0.061 | 0.84 | 0.061 | 0.74 | 0.049 | 0.74 | 0.049 |
| u8 | 0.95 | 0.015 | 0.84 | 0.062 | 0.82 | 0.031 | 0.76 | 0.042 |
| u9 | 0.88 | 0.042 | 0.87 | 0.047 | 0.81 | 0.031 | 0.83 | 0.028 |
| u10 | 0.96 | 0.013 | 0.85 | 0.055 | 0.95 | 0.007 | 0.74 | 0.049 |

composed of just one CPU. However, these differences vanish and even get reversed when we move to actually parallel computing platforms (see columns "4 CPUs" and "8 CPUs" in Table 7.5). In general, execution times are very similar and differences are not statistically significant in many cases (see "-" symbols in columns KW−2).

Two metrics have been used in order to enrich our understanding of the effects of parallelism on the parallel algorithms of this work: the parallel efficiency ($\eta$) and the serial fraction ($sf$) [89]. If we consider that $N$ is the number of processors and $s_N$ is the speedup ($s_N = \bar{t}_{1\ CPU}/\bar{t}_{N\ CPUs}$), the two metrics can be defined as:

$$\eta = \frac{s_N}{N} = \frac{\frac{\bar{t}_{1\ CPU}}{\bar{t}_{N\ CPUs}}}{N} , \tag{7.11}$$

$$sf = \frac{\frac{1}{s_N} - \frac{1}{N}}{1 - \frac{1}{N}} . \tag{7.12}$$

Table 7.6 includes the resulting values of the metrics. Values of the parallel efficiency show that all the parallel versions of GA and SS are able to profit quite well from the parallel computing platform. Averaging over all the problems, pGA-4 gets an $\eta$ value of 0.87, while pSS-4 obtains 0.90. If we consider now the parallelization based on 8 islands, pGA-8 reaches a parallel efficiency of 0.79 whereas pSS-8 achieves a value of 0.85 (also averaging over all the problems). From these average values we can conclude that pSS algorithms profit better from the parallel platform than pGAs although the latter ones are faster in terms of absolute running times.

**Table 7.7** Hybrid GA results for structured and unstructured problems.

| Prob. | Seq. hGA | PhGA-4 | PhGA-8 | $KW_3$ |
|---|---|---|---|---|
| s1 | 913 | 870 | **867** | + |
| s2 | 959 | **912** | 920 | + |
| s3 | 1056 | 1021 | **1001** | + |
| s4 | 1007 | 999 | **997** | + |
| s5 | 1103 | **998** | 1006 | + |
| s6 | 1084 | 1040 | **1034** | + |
| s7 | 954 | **933** | **933** | + |
| s8 | 1295 | 1077 | **1067** | + |
| s9 | 985 | **948** | 952 | + |
| s10 | 934 | 903 | **891** | + |
| u1 | 1375 | 1361 | **1280** | + |
| u2 | 1193 | 1176 | **1098** | + |
| u3 | 1509 | 1204 | **1187** | + |
| u4 | 1670 | 1342 | **1286** | + |
| u5 | 1189 | 1173 | **1170** | + |
| u6 | 1193 | 1174 | **1128** | + |
| u7 | 1288 | 1163 | **1106** | + |
| u8 | 1076 | 1055 | **1041** | + |
| u9 | 927 | 894 | **883** | + |
| u10 | 1205 | 1086 | **998** | + |

If we compare the parallel efficiency of the algorithms when the number of processors increases, it can be seen in Table 7.6 that there is a reduction in the values of this metric and the average values presented previously also support this claim. Here, the serial fraction metric plays an important role. If the values of this metric remain almost constant when using a different number of processors in a parallel algorithm, it allows us to conclude that the loss of efficiency is because of the limited parallelism of the model itself and not because our implementation. For example in the instance s5 with pGAs: the parallel efficiency is reduced by 15% (from 0.78 in pGA-4 down to 0.63 in pGA-8) while the serial fraction is almost the same (0.089 in pGA-4 against 0.080 in pGA-8), confirming the previous hypothesis that the loss of efficiency is due to the implementation parallel model.

## 7.4.4 A Parallel Hybrid GA

We suggested in Section 7.4.2 that the better workforce planning performance reached by the SS algorithm did lie in the improvement operator used. In order to further investigate this fact, we have developed a new hybrid genetic algorithm (hGA) in which the SS improvement method has been incorporated into the GA main loop as an evolutionary operator. Specifically, the local search algorithm is applied just after the recombination and mutation operators by using a predefined probability which has been set up to $\rho_h = \frac{10}{population\_size}$. The experiments conducted in the following go towards validating the previous claim, so neither the parallel versions are executed on a single processor nor the computational times are presented: only the workforce planning performance is studied for the new hGA, the GA, and the SS. The results of the new hGA are included in Table 7.7. The stopping

**Table 7.8** Average results for structured and unstructured problems.

| Problems | s1 - s10 | | | u1 - u10 | | |
|---|---|---|---|---|---|---|
| Algorithm | GA | SS | hGA | GA | SS | hGA |
| Sequential | 0.9989 | 0.9405 | 0.9202 | 0.9735 | 0.9744 | 0.9309 |
| 4 Islands | 0.8842 | 0.8743 | 0.8691 | 0.8877 | 0.8590 | 0.8626 |
| 8 Islands | 0.8771 | 0.8658 | 0.8641 | 0.8711 | 0.8284 | 0.8294 |

condition of all these algorithms is the same as in the previous experiments: reaching a predefined number of function evaluations. In the parallel versions, the hGA also follows the island model described in Section 7.2.6.

Let us start analyzing the workforce planning performance for the structured instances. As it happened with the GA, the sequential hGA is always outperformed by its two parallel versions. These performance improvements range from 1% in s4 to 21.3% in s8 (6.46% on average over all the structured instances). Concerning PhGA-4 and PhGA-8, the latter computes the highest performance workforce planning in 6 out of the 10 structured instances, but now differences are smaller (0.75% on average). The diversity introduced by the parallel models is clearly the responsible for this results. The explanation for this claim concerns the loss of diversity provoked by the newly introduced improvement operator and, consequently, the parallel hGA models counteract to some extent the increasingly chance of getting trapped in a local minimum.

In the case of the unstructured instances, the previous claims are even more evident. Here, PhGA-8 gets the best planning performance in 10 out of the 10 instances, and with statistical confidence ("+" symbols in the last column). With respect to the sequential hGA, a noticeable reduction has been reached in u3 and u4 (27.12% and 29.86%, respectively). Averaging over all the instances, PhGA-8 has been able to reduce the planning costs a percentage of 12.6%. Comparing the two parallel versions, differences now are not that smaller, e.g. reaching almost 9% in u10 (from 1086 down to 998). We can therefore conclude that diversity is even more decisive when solving the unstructured version of the WPP with the new hybrid GA.

In order to compare this new proposal against the previously presented algorithms we have followed the same approach as in Section 7.4.2: we have normalized with respect to all the maximum (worst) values, thus avoiding scaling problems. The results are presented in Table 7.8. Note that the values for GA and SS are different from those included in Table 7.4 since the values used for normalization have changed.

If we have a look at the sequential versions of the three algorithms, it can be noticed that hGA outperforms GA and SS in both structured and unstructured instances. This also holds for the parallel models with 4 island in the structured instances. However, the pSS is the best parallel algorithm with 4 islands when solving the unstructured instances (an average value of 0.8590 against 0.8877 of GA and 0.8626 of hGA). The resulting values of the normalized workforce planning performance in the parallel models with 8 islands keep the same behavior: parallel hGA improves upon both parallel GA and parallel SS in the structured instances whereas SS is the best approach in the unstructured ones. The point here is that the differences are tighter. We can conclude that the hGA can profit from using the improvement operator because it always outperforms the GA approach. Concerning SS, hGA is able to always reach improved workforce plannings in the sequential case and in the structured instances.

Summarizing, we can conclude that hGA is the best algorithms among all the sequential algorithms used. Concerning parallel versions, a clear conclusion is that parallel versions always outperforms serial ones. Among the parallel methods, pGA is the worst while the pSS achieves the best accuracy, obtaining a slightly better performance than phGA.

## 7.5 Conclusions

In this chapter we have addressed and solved a workforce planning problem. To achieve this goal we have used two metaheuristics: a parallel GA and a parallel SS. The development of these parallel versions of a genetic algorithm and a parallel scatter search aims at tackling problems of realist size.

The conclusions of this work can be summarized attending to different criteria. Firstly, as it was expected, the parallel versions of the methods have reached an important reduction of the execution time with respect to the serial ones. In fact, our parallel implementations have obtained a very good speedup (nearly linear). In several instances, we have noticed a moderate loss of efficiency when increasing the number of processor from four to eight. But this loss of efficiency is mainly due to the limited parallelism of the program, since the variation in the serial fraction was negligible.

Secondly, we have observed that the parallelism did not only allow to reduce the execution time but it also allowed to improve the quality of the solutions. Even when the parallel algorithms were executed in a single processor, they outperformed the serial one, proving clearly that the serial and the parallel methods are different algorithms with different behaviors.

Thirdly, we have noticed that SS results outperformed GA ones for both kind of instances. The search scheme of the SS seems to be more appropriate to the WPP than the GA one. We have studied if the improvement operator used by SS is beneficial to this problem, and demonstrated that a hybridization of GA with this local search mechanism provokes an improvement in the quality of the solutions. This hybridization has allowed to ameliorate the performance of the "pure" GA in both kind of instances, whereas it was only able to outperform GA in the structured ones and SS only in the serial version.

As future work, a potential line of research is to apply these techniques to tackle instances ten times (or more) larger than those solved here (i.e., $n \approx m \approx 200$). This is a really challenging set of instances for new works in this area.

8

# Parallel GAs in Bioinformatics: Assembling DNA Fragments

*We used to think that our future was in the stars, now we know it is in our genes.*

*James Watson (1928 - ) - American scientist*

In this chapter, we deal with a quite recent and important domain in Computer Science: Bioinformatics. In concrete, we solve the DNA fragment assembly problem using a parallel GA. DNA fragment assembly is a technique that attempts to reconstruct the original DNA sequence from a large number of fragments, each one being several hundred base-pairs (bps) long. The DNA fragment assembly is needed because current technology, such as gel electrophoresis, cannot directly and accurately sequence DNA molecules longer than a few thousands bases. However, most real genomes are much longer. For example, a human DNA is about 3.2 billion bases in length and cannot be read at once.

The following technique was developed to deal with this limitation. First, the DNA molecule is amplified, that is, many copies of the molecule are created. The molecules are then cut at random sites to obtain fragments that are short enough to be sequenced directly. The overlapping fragments are then assembled back into the original DNA molecule.

The assembly problem is therefore a combinatorial optimization problem that, even in the absence of noise, is NP-hard: given $k$ fragments, there are $2^k k!$ possible combinations. This chapter reports on the design and implementation of a parallel distributed genetic algorithm to tackle the DNA fragment assembly problem. The pGA will then act as a DNA assembler, a hot topic in present research [182, 183].

The remainder of this chapter is organized as follows. In the next section, we present background information about the DNA fragment assembly problem. Section 2 presents a brief review of existing assemblers. In Section 3, we discuss the operators, fitness functions [184], and how to design and implement a parallel genetic algorithm (GA) for the DNA fragment assembly problem. We analyze the results of our experiments in Section 4. We end this chapter by giving our final thoughts and conclusions in Section 5.

G. Luque and E. Alba: Parallel Genetic Algorithms, SCI 367, pp. 135–147.
springerlink.com 

## 8.1 The Work of a DNA Fragment Assembler

We start this section by giving a vivid analogy to the fragment assembly problem: "Imagine several copies of a book cut by scissors into thousands of pieces, say 10 millions. Each copy is cut in an individual way such that a piece from one copy may overlap a piece from another copy. Assume one million pieces get lost and the remaining nine million are splashed with ink: try to recover the original book." [185]. We can think of the DNA target sequence as being the original text and the DNA fragments are the pieces cut out from the book. To further understand the problem, we need to know the following basic terminology:

- **Fragment:** A short sequence of DNA with length up to 1000 bps.
- **Shotgun data:** A set of fragments.
- **Prefix:** A substring comprising the first $n$ characters of fragment $f$.
- **Suffix:** A substring comprising the last $n$ characters of fragment $f$.
- **Overlap:** Common sequence between the suffix of one fragment and the prefix of another fragment.
- **Layout:** An alignment of collection of fragments based on the overlap order.
- **Contig:** A layout consisting of contiguous overlapping fragments.
- **Consensus:** A sequence derived from the layout by taking the majority vote for each column (base) of the layout.

To measure the quality of a consensus, we can look at the distribution of the coverage. Coverage at a base position is defined as the number of fragments at that position. It is a measure of the redundancy of the fragment data. It denotes the number of fragments, on average, in which a given nucleotide in the target DNA is expected to appear. It is computed as the number of bases read from fragments over the length of the target DNA [186].

$$Coverage = \frac{\sum_{i=1}^{n} length\ of\ the\ fragment\ i}{target\ sequence\ length} \tag{8.1}$$

where $n$ is the number of fragments. TIGR uses the coverage metric to ensure the correctness of the assembly result. The coverage usually ranges from 6 to 10 [187]. The higher the coverage, the fewer the gaps are expected, and the better the expected result (easier problem).

### 8.1.1 DNA Sequencing Process

To determine the function of specific genes, scientists have learned to read the sequence of nucleotides comprising a DNA sequence in a process called DNA sequencing. The fragment assembly starts with breaking the given DNA sequence into small fragments. To do that, multiple exact copies of the original DNA sequence are made. Each copy is then cut into short

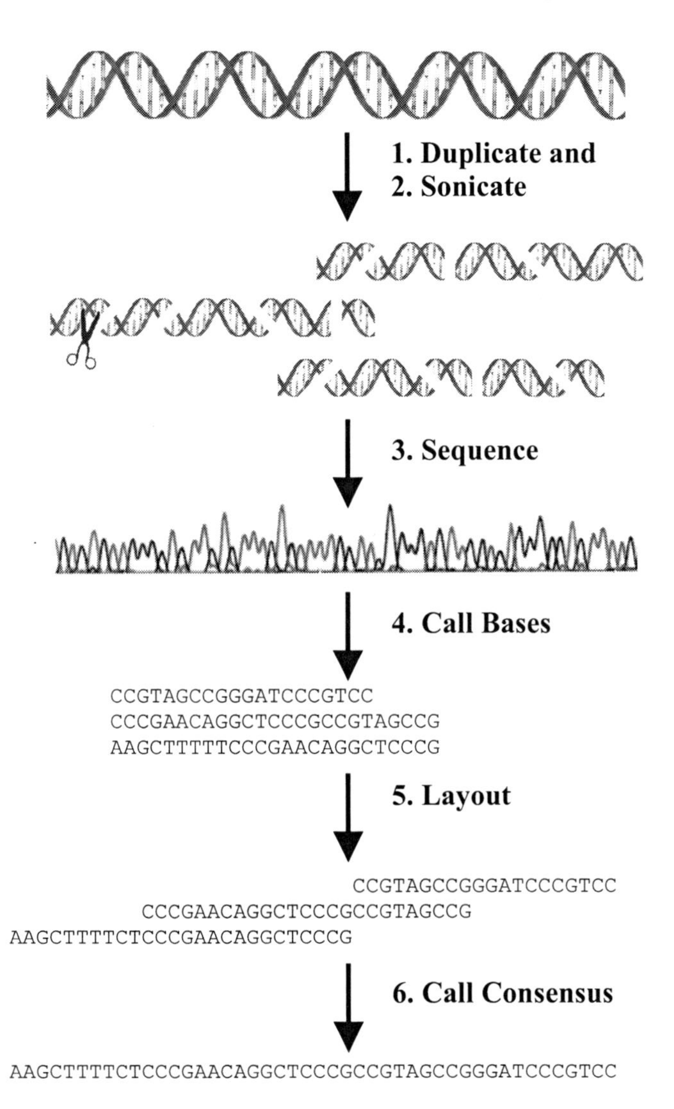


**Fig. 8.1** Graphical representation of DNA sequencing and assembly [188].

fragments at random positions. These are the first three steps depicted in Fig. 8.1 and they take place in the laboratory. After the fragment set is obtained, the standard assemble approach is followed in this order: overlap, layout, and then consensus. To ensure that fragments overlap, the reading of fragments continues until the coverage is satisfied. These steps are the last three steps in Fig. 8.1. In what follows, we give a brief description of each of the three phases, namely overlap, layout, and consensus.

**Overlap Phase** - *Finding the overlapping fragments.*

Usually, overlaps are available offline before starting this work. This phase consists in finding the best or longest match between the suffix of one sequence and the prefix of another. In this step, we compare all possible pairs of fragments to determine their similarity. Usually, the dynamic programming algorithm applied to semiglobal alignment is used in this step. The intuition behind finding the pairwise overlap is that fragments with a significant overlap score are very likely next to each other in the target sequence.

**Layout Phase** - *Finding the order of fragments based on the computed similarity score.*

This is the most difficult step because it is hard to tell the true overlap due to the following challenges:

1. Unknown orientation: After the original sequence is cut into many fragments, the orientation is lost. The sequence can be read in either 5' to 3' or 3' to 5'. One does not know which strand should be selected. If one fragment does not overlap with another, it is still possible that its reverse complement might have such an overlap.
2. Base call errors: There are three types of base call errors: substitution, insertion, and deletion. They occur due to experimental errors in the electrophoresis procedure. Errors affect the detection of fragment overlaps. Hence, the consensus determination requires multiple alignments in high covered regions.
3. Incomplete coverage: It happens when the algorithm is not able to assemble a given set of fragments into a single contig.
4. Repeated regions: Repeats are sequences that appear two or more times in the target DNA. Repeated regions have caused problems in many genome-sequencing projects, and none of the current assembly programs can handle them perfectly.
5. Chimeras and contamination: Chimeras arise when two fragments that are not adjacent or overlapping on the target molecule join together into one fragment. Contamination occurs due to the incomplete purification of the fragment from the vector DNA.

After the order is determined, the progressive alignment algorithm is applied to combine all the pairwise alignments obtained in the overlap phase.

**Consensus Phase** - *Deriving the DNA sequence from the layout.*

The most common technique used in this phase is to apply the majority rule (it assigns the base with more repetitions in each position) in building the consensus.

*Example:* We next give an example of the fragment assembly process.

Given a set of fragments {F1 = GTCAG, F2 = TCGGA, F3 = ATGTC, F4 = CGGATG}. First, we need to determine the overlap of each pair of the fragments by the using semiglobal alignment algorithm. Next, we determine the order of the fragments based on the overlap scores, which are calculated in the overlap phase. Suppose we have the following order: F2 F4 F3 F1. Then, the layout and the consensus for this example can be constructed as follows:

```
          F2 ->    TCGGA
          F4 ->     CGGATG
          F3 ->        ATGTC
          F1 ->          GTCAG
------------------------
Consensus -> TCGGATGTCAG
```

In this example, the resulting order allows to build a sequence having just one contig. Since finding the exact order takes a huge amount of time, a heuristic such as Genetic Algorithm can be applied in this step [184, 189, 190]. In the following section, we discuss several existing algorithms for the DNA fragment assembly problem. ∎

From a combinatorial optimization viewpoint, the whole process which results in the construction of the consensus sequence is similar to that of a tour in the Travelling Salesman Problem (TSP). This is because each fragment would have to be in a specific fragment ordering sequence in order for the formation of a consensus sequence to take place. The main difference between TSP and DNA fragment assembly is that there would *not* be a proper alignment between the first and the last fragments in the consensus sequence that is comparable to the connection between the first and the last cities in the TSP solution. Therefore, many equivalent solutions to TSP are thus nonequivalent in our context. Other important difference is that, while the ordering is the final solution to TSP, in our case, this ordering is only an intermediate step (final consensus is still needed) and then several different orderings can produce equivalent results. Other minor differences can be found between both problems due to the challenges described previously (unknown orientation, incomplete coverage, etc.).

## 8.2 Related Literature

Originally, the assembly of short fragments was done by hand, which is inefficient and error-prone. Hence, a lot of effort has been put into finding techniques to automate the sequence assembly. Over the past decade a number of fragment assembly packages have been developed and used to sequence different organisms. The most popular packages are PHRAP [191], TIGR assembler [192], STROLL [193], CAP3 [194], Celera assembler [195], EULER [185], and PALS [196]. These packages deal with the previously described challenges, but none of them solves them all. Each package automates fragment assembly using a variety of algorithms. The most popular techniques are greedy-based.

Also, some metaheuristic methods have been applied to this problem, like SA [197], GA [184, 197], and VNS [198]. These methods obtain quite accurate result for small/medium instances but they have some difficulties to solve real-world problem instances. Therefore, some enhanced techniques were proposed such as parallel [197], and hybrid methods [199].

## 8.3 The pGA DNA Assembler

In this section, we describe our parallel GA used to solve the DNA fragment assembly problem. First, we show the main details about the representation, fitness evaluation, and operators. Then, we discuss the parallel model used.

### 8.3.1 Solution Encoding

We use the permutation representation with integer number encoding. A permutation of integers represents a sequence of fragment numbers, where successive fragments overlap. A tentative solution in this representation requires a list of fragments assigned with a unique integer ID. For example, 8 fragments will need eight identifiers: 0, 1, 2, 3, 4, 5, 6, 7. The permutation representation requires special operators to make sure that we always get legal (feasible) solutions. In order to maintain a legal solution, the two conditions that must be satisfied are (1) all fragments must be presented in the ordering, and (2) no duplicate fragments are allowed in the ordering. For example, one possible ordering for 4 fragments is 3 0 2 1. It means that fragment 3 is at the first position and fragment 0 is at the second position, and so on.

### 8.3.2 Solution Evaluation

A fitness function is used to evaluate how good a particular solution is, i.e., to define the objective for the optimization process. It is applied to each individual in the population and it should guide the genetic algorithm towards the optimal solution. In the DNA fragment assembly problem, the fitness function measures the multiple sequences alignment quality and finds the best scoring alignment. Parsons, Forrest, and Burks mentioned two different fitness functions [184].

**Fitness function** $F1$ - it sums the overlap score ($w$) for adjacent fragments in a given solution. When this fitness function is used, the objective is to maximize such a score. It means that the best individual will have the highest score.

$$F1(l) = \sum_{i=0}^{n-2} w(f[i], f[i+1]) \tag{8.2}$$

**Fitness function** $F2$ - it not only sums the overlap score for adjacent fragments, but also sums the overlap score for all other possible pairs.

$$F2(l) = \sum_{i=0}^{n-1} \sum_{j=0}^{n-1} |i-j| \times w(f[i], f[j]) \tag{8.3}$$

This fitness function penalizes solutions in which strong overlaps occur between non-adjacent fragments in the layouts. When this fitness function is used, the objective is to minimize the overlap score. It means that the best individual will have the lowest score.

The overlap score in both $F1$ and $F2$ is computed using the semiglobal alignment algorithm.

### 8.3.3 Genetic Operators

In order to generate a new population, GA applies three operators: recombination, mutation, and selection/replacement. In the following paragraphs, we explain the operators used in this work.

*Recombination Operator*

Two or more parents are recombined (or crossed over) to produce one or more offspring. The purpose of this operator is to allow partial solutions to evolve in different individuals and then combine them to produce a better solution. It is implemented by running through the population and for each individual, deciding whether it should be selected for recombine using a parameter called *recombination rate* ($P_r$). A recombination rate of 1.0 indicates that all the selected individuals are used in the recombination operator. Thus, there are no survivors. However, empirical studies have shown that better results are achieved by a recombination rate between 0.65 and 0.85, which implies that the probability of an individual moving unchanged to the next generation ranges from 0.15 to 0.35.

For our experimental runs, we use the order-based recombination operator (OX) and the edge-recombination operator (ER). These operator were specifically designed for tackling problems with permutation representations.

The order-based operator first copies the fragment ID between two random positions in Parent1 into the offspring's corresponding positions. It then copies the rest of the fragments from Parent2 into the offspring in the relative order presented in Parent2. If the fragment ID is already present in the offspring, then we skip that fragment. The method preserves the feasibility of every string in the population.

Edge recombination preserves the adjacencies that are common to both parents. This operator is appropriate because a good fragment ordering consists of fragments that are related to each other by a similarity metric and should therefore be adjacent to one another. Parsons [190] defines edge recombination operator as follows:

1. Calculate the adjacencies.
2. Select the first position from one of the parents, call it $s$.
3. Select $s'$ in the following order until no fragments remain:
   a. $s'$ adjacent to $s$ is selected if it is shared by both parents.
   b. $s'$ that has more remaining adjacencies is selected.
   c. $s'$ is randomly selected if it has an equal number of remaining adjacencies.

*Mutation Operator*
This operator is used for the modification of single individuals. The reason we need a mutation operator is for the purpose of maintaining diversity in the population. Mutation is implemented by running through the whole population, and for each individual, deciding whether to select it for mutation or not, based on a parameter called *mutation rate* ($P_m$). For our experiments, we use the swap mutation operator. This operator randomly selects two positions from a permutation and then swaps the two fragment positions. Since this operator does not introduce any duplicate value in the permutation, the resulting solution is always feasible. Swap mutation operator is suitable for permutation problems like ordering fragments.

*Selection operator*
The purpose of the selection is to weed out the bad solutions. It requires a population as a parameter, processes the population using the fitness function, and returns a new population. The level of the selection pressure is very important. If the pressure is too low, convergence becomes very slow. If the pressure is too high, convergence will be premature to a local optimum.

In this work, we use ranking selection mechanism [200], in which the GA first sorts the individuals based on the fitness and then selects the individuals with the best fitness score until the specified population size is reached. Note that the population size will grow whenever a new offspring is produced by recombination or mutation because we consider a temporary population in every step made of the old population and the new one. The use of ranking selection is preferred over other selections such as fitness proportional selection [116].

### 8.3.4 The Parallel Approach

This section introduces the parallel models that we use in the experiments discussed in the next section.

A parallel GA (PGA) is an algorithm having multiple component GAs, regardless of their population structure (Chapter 2). A component GA is usually a traditional GA with a single population. Its algorithm is augmented with an additional phase of *communication* code so as to be able to disseminate its result and receive results from the other components [6].

Different parallel algorithms differ in the characteristics of their elementary heuristics and in the communication details. In this work, we have chosen a kind of decentralized distributed search because of its popularity and because it can be easily implemented in clusters of machines. In this parallel implementation, separate subpopulations evolve independently in a ring with sparse exchanges of a given number of individuals with a certain given frequency. The selection of the emigrants is through binary tournament [116] in the component genetic algorithms, and the arriving immigrants replace the worst ones in the population only if the new one is better than this current worst individuals.

## 8.4 Experimental Validation

In this section we analyze the behavior of our parallel GA. To measure the quality of its results, we will compare its performance with respect to other algorithms presented in the literature (GA, SA, and parallel SA) [197]. We have selected these methods since they represent both population and trajectory based metaheuristics.

Firstly, the target problem instance used is presented. Then, we show the parameterization of the methods and, finally, experiments performed are included at the end of this section.

### *8.4.1 Target Problem Instances*

A target sequence with accession number BX842596 (GI 38524243) was used in this work. It was obtained from the NCBI web site[1]. It is the sequence of a Neurospora crassa (common bread mold) BAC, and is 77,292 base pairs long. To test and analyze the performance of our algorithm, two problem instances were generated by GenFrag [201]. The first problem instance, 842596_4, contains 442 fragments with average fragment length of 708 bps and coverage 4, while the second one 842596_7, contains 773 fragments with average fragment length of 703 bps and coverage 7.

These instances are very hard, since they were generated from very long sequence using a small/medium value of coverage and a very restrictive cutoff. The combination of these parameters is known to produce very complex instances. For example, longer target sequences have been solved in the literature [194], however they used a higher coverage. The coverage measures the redundancy of the data: the higher the coverage, the easier the problem. The cutoff value is the minimum overlap score between two adjacent fragments required to join them in a single fragment. The cutoff, which we have set

[1] `http://www.ncbi.nlm.nih.gov/`

to thirty (a very high value), provides one filter for spurious overlaps introduced by experimental error. Instances with these features have been only solved adequately when target sequences vary from 20k to 50k base pairs [184, 202], while here, we are considering instances with 77k base pairs. We evaluated each assembly result in terms of the number of contigs assembled and the percentage similarity of assembled regions with the target sequence. Since we obtain fragments from a known target sequence, we can compare our assembled consensus sequence with the original target.

### 8.4.2 Parameterization

To solve this problem, we use a sequential GA, a sequential SA, several distributed GAs (having 2, 4, and 8 islands), and several parallel SAs (having 2, 4, and 8 components).

Since the results of GA and SA vary depending on the different parameter settings, we previously performed a complete analysis to study how the parameters affect the results and the performance of algorithms. From these previous analyses, we conclude that the best settings for our problem instances of the fragment assembly problem is a population size of 512 individuals, with F1 as fitness function, OR as crossover operator (with probability 1.0), and with a swap mutation operator (with probability 0.3). The migration in dGAs occurs in a unidirectional ring manner, sending one single randomly chosen individual to the neighbor sub-population. The target population incorporates this individual only if it is better than its present worst solution. The migration step is performed every 20 iterations in every island in an asynchronous way. For SA we use a Markov chain of length *total number evaluations*/100 and a proportional cooling scheme with a decreasing factor of 0.99. Each component SA exchanges, every 5000 evaluations, the best solution found (*cooperation* phase) with its neighbor SA in the ring. Because of the stochastic nature of the algorithms, we perform 30 independent runs of each test to gain sufficient experimental data. A summary of the conditions for our experimentation is found in Table 8.1.

### 8.4.3 Analysis of Results

Table 8.2 shows all the results and performance metrics for all data instances and algorithms described in this chapter. The table shows the fitness of the best solution obtained ($b$), the average fitness found ($f$), average number of evaluations ($e$), and average time in seconds ($t$). The statistical test for the time column are always positive, i.e., all times are significantly different from each other. The statistical test for the evaluation column reveals that there exists a significant difference in the number of visited points between the sequential algorithm and the parallel versions, i.e., the parallel model allows to change the behavior of the method, and for this problem, that change is beneficial: a significant reduction in the effort for computing a solution.

**Table 8.1** Parameter Settings.

| Genetic Algorithms | |
|---|---|
| Independent runs | 30 |
| Popsize | 512 |
| Fitness function | F1 |
| Crossover | OR (1.0) |
| Mutation | Swap (0.3) |
| Cutoff | 30 |
| Migration frequency | 20 iterations |
| Migration rate | 1 |
| **Simulated Annealing** | |
| Independent runs | 30 |
| Fitness function | F1 |
| Move operator | Swap |
| Cutoff | 30 |
| Markov chain length | $\frac{total\ number\ evaluations}{100}$ |
| Cooperation phase | 5000 evaluations |

**Table 8.2** Results for the two problem instances.

| | | **842596_4** | | | | **842596_7** | | | |
|---|---|---|---|---|---|---|---|---|---|
| Algorithm | | b | f | e | t | b | f | e | t |
| GA | Sequential | 33071 | 27500 | 810274 | 74 | 78624 | 67223 | 502167 | 120 |
| | LAN $n=2$ | **107148** | **88653** | 733909 | 36 | 156969 | 116605 | 611694 | 85 |
| | LAN $n=4$ | 88389 | 74048 | 726830 | 18 | 158021 | **120234** | 577873 | 48 |
| | LAN $n=8$ | 66588 | 58556 | 539296 | 8.5 | **159654** | 119735 | 581979 | 27 |
| SA | Sequential | 41893 | 36303 | 952373 | 58 | 81624 | 76303 | 897627 | 91 |
| | LAN $n=2$ | 79406 | 74238 | 929146 | 27 | 120357 | 101374 | 782915 | 51 |
| | LAN $n=4$ | 83820 | 76937 | 931673 | 15 | 134917 | 118724 | 819218 | 28 |
| | LAN $n=8$ | 80914 | 75193 | 915393 | 7 | 126934 | 116290 | 825397 | 15 |

Let us now discuss some of the results found in the table. First, for both instances, it is clear that the parallel version outperforms the serial one. Both, parallel GA and parallel SA, yield better fitness values and are faster than the sequential GA and SA, respectively. The parallel GA solutions are more accurate than the parallel SA ones, while the sequential version of SA outperforms the sequential GA version. The SA execution is faster than the GA one, because the GA executes an additional time-consuming operator (order-based crossover). Let us now go in deeper details on these claims.

For the first problem instance, the parallel GAs sampled less points in the search space than the serial GA, while for the second instance the serial algorithm is mostly similar in the required effort with respect to the parallel ones. The parallel SAs reduce the number of evaluations in both instances.

Increasing of the number of islands or components (and CPUs) results in a reduction in search time, but it does not lead to a better fitness value.

**Table 8.3** Speed-up.

| Algorithms | | 842596_4 | | | 842596_7 | | |
|---|---|---|---|---|---|---|---|
| | | $n$ CPUs | 1 CPU | Speedup | $n$ CPUs | 1 CPU | Speedup |
| GA | $n = 2$ | 36.21 | 72.07 | 1.99 | 85.37 | 160.15 | 1.87 |
| | $n = 4$ | 18.32 | 72.13 | 3.93 | 47.78 | 168.20 | 3.52 |
| | $n = 8$ | 8.52 | 64.41 | 7.56 | 26.81 | 172.13 | 6.42 |
| SA | $n = 2$ | 27.41 | 52.35 | 1.92 | 51.33 | 84.69 | 1.65 |
| | $n = 4$ | 14.89 | 55.12 | 3.74 | 27.76 | 89.38 | 3.22 |
| | $n = 8$ | 7.10 | 51.75 | 7.29 | 14.63 | 90.26 | 6.17 |

This is a negative behavior of the GA. For the second problem instance, the average fitness was improved by a larger number of CPUs. However, for the first problem instance, we observed a reduction in the fitness value as we increased the number of CPUs. This counterintuitive result clearly states that each instance has a different number of optimum from the point of view of the accuracy.

In the parallel GA, the best tradeoff is for two islands (n = 2) for the two instances, since this value yields a high fitness at an affordable cost and time. In the case of parallel SA, the optimum configuration is obtained when it uses four components.

Table 8.3 gives the speed-up results. As it can be seen in the table, we always obtain an almost linear speedup for the first problem instance. This is a very good result. For the second instance we also have a good speedup with a low number of CPUs (two and four components); eight islands make the efficiency decrease to a moderate speedup (6.42 for parallel GA and 6.22 for parallel SA). The speed-ups obtained by parallel GA are always better than the parallel SA ones.

**Table 8.4** Final number of best contigs.

| Algorithms | | 842596_4 | 842596_7 |
|---|---|---|---|
| Sequential | | 5 | 4 |
| GA LAN | n = 2 | 3 | 2 |
| | n = 4 | 4 | **1** |
| | n = 8 | 4 | 2 |
| Average | | 4 | **2.25** |
| Sequential | | 5 | 4 |
| SA LAN | n = 2 | 5 | 3 |
| | n = 4 | 4 | **2** |
| | n = 8 | 5 | **2** |
| Average | | 4.75 | **2.75** |

Finally, Table 8.4 shows the best global number of contigs computed in every case. This value is used as a high-level criterion (the more important one) to judge the whole quality of the results since, as we said before, it is difficult to capture the dynamics of the problem into a mathematical (fitness) function. These values are computed by applying a final step of refinement with a greedy heuristic commonly used in this application [202]. We have found that in some (extreme) cases it is even possible that a solution with a better fitness than other one generates a larger number of contigs (worse solution). This is the reason for still needing research to get a more accurate mapping from fitness to contig number. The values of this table confirm again that all the parallel versions outperform the serial versions, thus advising the utilization of parallel metaheuristics for this application in the future. Also, GA outperforms SA on average for sequential and parallel versions, since in the first problem, average contig is 4 for GA versus 4.75 of SA, and for the second is 2.25 for GA versus a higher 2.75 of SA.

## 8.5 Conclusions

The DNA fragment assembly is a very complex problem in Bioinformatics. Since the problem is NP-hard, the optimal solution is impossible to find for real cases, except for very small problem instances. Hence, computational techniques of affordable complexity such as metaheuristics are needed.

The sequential Genetic Algorithm we used here solves the DNA fragment assembly problem by applying a set of genetic operators and parameter settings, but does take a large amount of time for problem instances that are over 15K base pairs. Our distributed version has taken care of this shortcoming. Our test data are over 77K base pairs long. We are encouraged by the results obtained by our parallel algorithms not only because of their low waiting times, but also because of their high accuracy in computing solutions of even just 1 contig (the optimal solution in Nature). This is noticeable since it is far from trivial to compute optimal solutions for real-world instances of this problem.

We also faced to solve the DNA fragment assembly with a local search method, the simulated annealing. In this case, the results confirm that the parallelism helps to improve solutions and to reduce runtime also for this metaheuristic. On its own, separate SA executions are faster than GA ones, but generally, they lead to a worse fitness value.

A future open research lines, we can advise to analyze other kinds of parallel models created as extensions of the canonical skeletons used in this chapter. To curb the problem of premature convergence for example, we could propose a restart technique in the islands. Another interesting point of research would be to incorporate different algorithms in the islands (heterogeneous pGA), such as greedy or problem-dependent heuristics, and to study the effects this could have on the observed performance. Other interesting research line is testing the algorithmic approach with more realistic instances (for example, noisy cases [203]).

# A

# The MALLBA Library

MALLBA[1] [64, 204] is an effort to develop an integrated library of skeletons for combinatorial optimization including exact, heuristic and hybrid techniques. Sequential and parallel execution environments are supported in a user-friendly and, at the same time, efficient manner. Concerning parallel environments, both Local Area Networks (LANs) of workstations and Wide Area Networks (WANs) are considered.

All algorithms in the MALLBA library are implemented as *software skeletons* (similar to strategy pattern [205]) with a common internal and public interface. A skeleton is an algorithmic unit that, in a template-like manner, implements a generic algorithm. The algorithm will be made particular to solve a concrete problem by fulfilling the requirements specified in its interface. This permits fast prototyping and transparent access to parallel platforms. In the MALLBA library, every skeleton implements a resolution technique for optimization, taken from the fields of exact, heuristic and hybrid optimization.

All this software has been developed in C++. This language provides a high-level oriented-object set of services and, at the same time, it generates efficient executable files, what is an important issue for an optimization library. The skeleton design in MALLBA is based on the separation of two concepts: the features of the problem to be solved and the general resolution technique to be used. While the particular features related to the problem must be given by the user, the technique and the knowledge needed to parallelize the execution of the resolution technique is implemented in the skeleton itself, and is completely provided by the library. Thus the user does not program neither the resolution technique nor its parallelization. It is very common that the problem is represented by a complex function to be optimized and the details on how to manipulate tentative solutions (e.g. merge, cut, or interpret parts of them). Basically, the resolution technique is the algorithm defining

[1] Source code and examples can be found at
`http://neo.lcc.uma.es/software/mallba/index.php`

the steps to proceed to the optimization of the problem. Almost every optimization technique exhibits a traditional three stage process, namely: (1) generating initial solutions (2) an improvement loop and (3) testing a stop condition. The way in which different skeletons do this work is really different and varied in the actual spectrum of optimization search and learning.

Skeletons are implemented as a set of *required* and *provided* C++ classes which represent object abstractions of the entities participating in the resolution technique (see Fig. A.1). The *provided* classes implement internal aspects of the skeleton in a problem-independent way. This internal set of classes is called the *Solver part* of the skeleton. In general, for each algorithmic technique several sequential resolution patterns are provided, all of them grouped in the class `Solver_Seq` (for example, the iterative and recursive patterns showed in Fig. A.1). The parallel patterns are grouped in the classes `Solver_Lan` and `Solver_Wan`. In the figure are depicted resolution patterns which use the master-slave paradigm, independent runs, and replicate data. Those classes are completely implemented and provided in the respective skeletons. The *required* classes specify information related to the problem. For the whole skeleton to work, it is required that these classes get completed with problem-dependent information. This conceptual separation allows us to define required classes with a fixed interface but without an implementation, so that provided classes can use required classes in a generic way. The fulfillment of the required classes would make the skeleton applicable to virtually any problem specified.

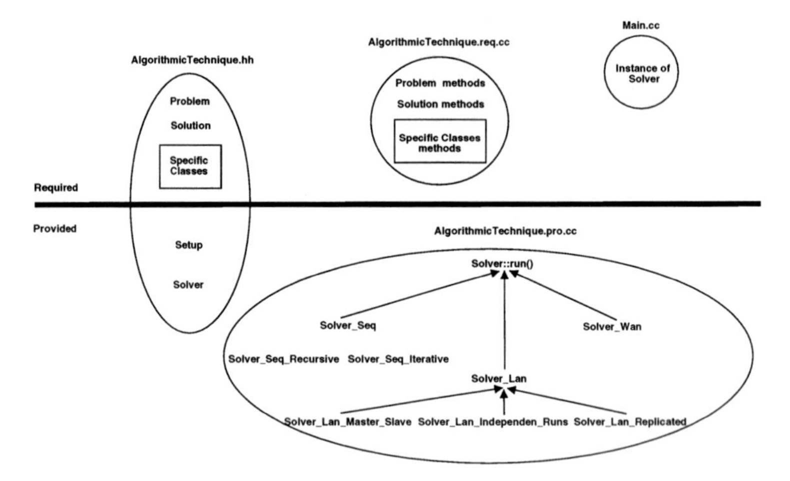


**Fig. A.1** Architecture of a MALLBA skeleton. The horizontal line stands for the separation between the C++ classes the user must fulfill (upper part) and the classes that the skeleton already includes in a fully operational form (lower part).

The fact that the user of a MALLBA skeleton only needs to implement the particular features of the problem to be solved, i.e., to fill in the required classes with an specific problem-dependent implementation, helps the creation of new programs with a very small effort.

Next, we will introduce and discuss the external interface of a skeleton. Two kinds of users will want to work with this interface: the user who wants to instantiate a new problem, and the user who wants to implement a new skeleton and incorporate it to the MALLBA library. Also in the next subsections, we will discuss the communication and the hybridization interfaces. The aim is to explain first what a final user must consider, then what a skeleton programmer needs to know about the parallel issues and, finally, how to merge skeletons to yield new optimization procedures. Descriptions from these different points of view are needed since the potential users and researchers using the library will have a different level of interaction with this software (see Fig. A.2), depending on his/her goal (e.g., to instantiate a problem, to change the communication layer, or to create new skeletons for new techniques).

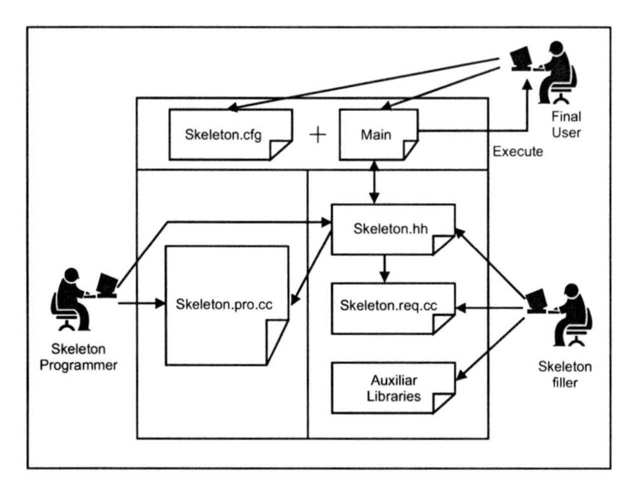


**Fig. A.2** Interaction of the different user types and the files conforming a MALLBA skeleton. Usually, the skeleton filler and the final user are the same.

## A.1 Skeleton Interfaces

From the user's point of view, two major aspects must be considered: the problem to be solved and the resolution technique to be used. The user will be responsible for adequately describing the former. As to the latter, rather complete descriptions are provided by the library. The user addresses these two aspects by selecting the skeleton and implementing its problem-dependent aspects. Later, since LAN and WAN implementations exist, the user will be able to execute the resulting program on sequential or parallel environments. Notice however that the same unique description made by the user will be used in any of the environments. Fig. A.2 shows the interaction of different kinds of users (namely, final user, programmer and internal filler) with the different parts conforming a skeleton. Although there are three profiles, this does not mean that it is needed three different users to use MALLBA library. The same user can take care of the tasks at any profile level.

MALLBA already includes a large set of solvers ready for utilization. Extending them is quite direct, and creating new solvers is conceptually guided by the class hierarchy design. Parts of existing skeletons can be easily reused to construct new ones. Each skeleton could have its own configuration file to avoid recompilation when parameters change.

Apart from some illustrative examples, MALLBA does not contain any complete implementation of specific problems. On the contrary, it provides the generic code the user has to customize. In this way, a single implementation –abstract yet efficient– can be reutilized in different contexts. The user do not need to have a deep knowledge about parallelism or distributed computing; these aspects are already included in the library. Let us get deeper in our understanding of the provided and required C++ classes conforming a skeleton.

*Provided Classes:*

They implement internal aspects of the skeleton in a problem-independent way. The most important *provided* classes are `Solver` (the algorithm) and `SetUpParams` (setup parameters). Provided classes are implemented in the files having the `.pro.cc` extension (see Fig. A.2).

*Required Classes:*

They specify information related to the problem. Each skeleton includes the `Problem` and `Solution` required classes that encapsulate the problem-dependent entities needed by the solver method. Depending on the skeleton other classes may be required. Required classes are implemented in the files having the `.req.cc` extension (see Fig. A.2).

## A.2 Communication Interface

Providing a parallel platform has been one of the central objectives in MALLBA. Local networks of computers are nowadays a very cheap and popular choice in laboratories and departments. Moreover, the available computational power of Internet is allowing the interconnection of these local networks, offering a plethora of possibilities for exploiting these resources.

To this end (i.e., using MALLBA onto a network of computers), it is necessary to have a communication mechanism allowing executing skeletons both in LAN and WAN. Since these skeletons are implemented in a high-level language, it is desirable for this communication mechanism to be also high-level;

besides, some other network services could be needed in the future would be needed such as the management of parallel processes.

The needed set of services is generically termed *middleware*, and it is responsible for all basic communication facilities. Several steps were followed to construct this system: first, related existing systems were studied and evaluated; then, a service proposal was elaborated; finally, the middleware was implemented in C++.

The detailed review of existing tools included both systems based in the message-passing paradigm and systems for the execution and management of distributed objects and programs. PVM, MPI, Java RMI, CORBA and Globus, as well as some other specific libraries, were evaluated [206]. The main conclusion was the need for an own system, adapted to the necessities of this library, but based on an efficient standard, capable of being valid in the future.

Meeting all these criteria can be, almost exclusively, possible by using MPI as the base for developing a communication library. Efficiency was a major goal in this work, and hence this decision; besides, MPI (in both MPICH and LAM/MPI, the two well-known implementations of the standard) is becoming increasingly popular, and has been successfully integrated in new promising systems such as Globus.

Although there is no theoretical drawback in using MPI directly, a light middleware layer termed `NetStream` (see Fig. A.3) was developed. With this tool, a MALLBA programmer can avoid the large list of parameters and interact with the network in the form of *stream modificators*, that allows advanced input/output operations to look like basic data exchanges with streams. By using `<<` and `>>` operators the programmer can develop LAN and WAN skeletons by feeding data and net operations in an easy way.

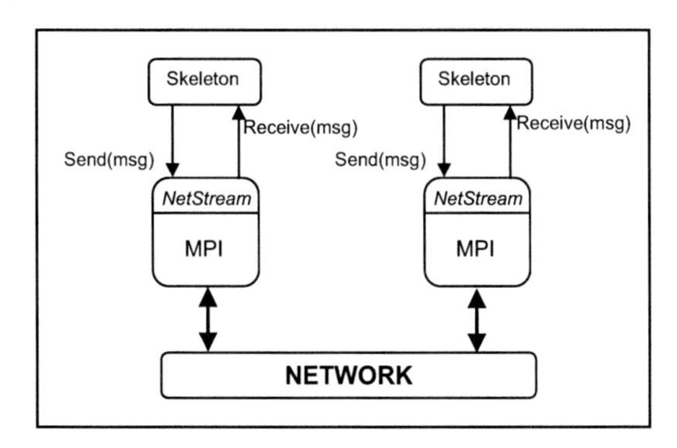


**Fig. A.3** NetStream communication layer on top of MPI.

Then, `NetStream` allows skeletons exchanging data structures efficiently, keeping a high abstraction level and ease of use. For this latter purpose, the number of parameters in the resulting methods has been minimized, and a large number of services has been implemented. These services can be classified into two groups: basic services, and advanced services. Among the basic ones we can mention:

- **Send-Reception of primitive data types:** `int`, `double`, `char`, strings, etc. both in raw format and packed (for efficiency purposes when used on a WAN). This can be done using input/output streams from/to the network.
- **Synchronization services:** barriers, broadcasts, checking for pending messages, etc. As in the C++ standard, these services are available by means of manipulators, i.e., methods that alter the behavior of a stream, feeding it as if they were data.
- **Basic management of parallel processes:** querying a process ID or the number of processes, establishing and retrieving the IDs of processes at the ends of a stream, etc.
- **Miscellaneous:** starting and stopping the system (using static class methods (singleton pattern [205]) rather than instances methods), etc.

Among the advanced services implemented we can cite the following:

- **Management of groups of processes:** this allows skeletons to be arranged in parallel optimization regions. Available methods allow manipulating communicators and intercommunicators between groups, in the MPI sense. This organization could be important for certain distributed algorithms, especially in the case of hybrid algorithms.
- **Services to acquire the on-line state of the network:** this C++ methods are provided to allow working with a model of both communication links and the state of machines involved in the execution, all this under a real time basis, during the run of a skeleton. Basically, these services endow the skeleton programmer with C++ methods to check the delays in any link of the LAN or WAN for different packet sizes, plus the error rate (noise) in the link, and the load of a workstation in the net. Furthermore, independent clients in C, C++ and Java have been developed in addition to the mentioned one in order to make `NetStream` a stand alone communication layer for optimization and other applications at a minimum complexity and overhead.

All these services provide high-level programming and will ease taking on-line decisions in WAN algorithms, although we are still at the stage of developing "intelligent" algorithms that use this information to perform a more efficient search.

## A.3 Hybridization Interface

In this section we discuss the mechanisms available in MALLBA to foster combinations of skeletons in the quest for more efficient and accurate solvers. This raises the question of constructing efficient *hybrid algorithms*. In its broadest sense, hybridization [130] refers to the inclusion of problem-dependent knowledge in a general search algorithm in one of two ways [207]:

- **Strong hybridization:** problem knowledge is included as specific non-conventional problem-dependent representations and/or operators.
- **Weak hybridization:** several algorithms are combined in some manner to yield the new hybrid algorithm.

The term "hybridization" has been used with diverse meanings in different contexts. Here, we refer to the combination of different search algorithms (the so-called weak hybridization). As it has been shown in theory [208] and practice [130], hybridization is an essential mechanism for obtaining effective optimization algorithms for specific domains. For this reason, there exist in MALLBA several tools for building such hybrid skeletons. This contrasts with other optimization libraries that let the programmer alone when building new algorithms from existing ones.

Due to the fact that the algorithmic skeletons will be reutilized and combined both by MALLBA end-users and by specialists, it is necessary to specify in a standard and unified fashion the way these skeletons can interact. For this reason, the notion of a *skeleton state* was proposed. The state of skeleton is its connection point with the environment. By accessing this state, one can inspect the evolution of the search, and take decisions regarding future actions of the skeleton. For this latter reason, it is mandatory to have not only the means for inspecting the state, but also for modifying it on the fly. Thus, either a user or another skeleton can control the future direction of the search. This is done with independence of the actual implementation of the skeleton, a major advantage in any large-scale project.

The advantages of using a state is that combining skeletons has a low cost, despite the fact that uniformly defining the state is not trivial, and constitutes an open research topic [209]. This proposal is articulated around the two basic classes we mentioned before: `StateVariable` and `StateCenter`. The `StateVariable` class allows defining and manipulating any information element within the algorithm skeleton. The latter is the connection point that provides access to the state itself.

On the basis of these classes, constructing a hybrid algorithm is very easy: one has to simply specify the behavior pattern by means of the appropriate manipulation of the states of the skeletons being combined. As an example of the flexibility of this model we have developed meta-algorithms that define the way in which $n$ component skeletons interact each other. One simply has to specify the precise algorithm involved to instance this meta-algorithm to a concrete working hybrid skeleton; the behavior pattern is the same no matter which these component algorithms are. This philosophy of "make once instance many" can serve to produce different algorithms with the same underlying search pattern at a minimum cost.

## A.4 Additional Information about MALLBA

Now, we show some lists of information about the methods available in this library and its utilization to solve some problems. These lists are not complete, they only show the most representative items in each category.

### *Algorithms Already Implemented in MALLBA*

(Most of the listed algorithms have also parallel versions)

- Genetic algorithms (including distributed and cellular versions, and non-traditional techniques such as CHC and $\mu$CHC).
- Simulated annealing.
- Evolutionary strategies.
- Differential evolution.
- Particle swarm optimization (continuous and discrete versions).
- Ant colony systems (including a specialized version for dynamic environments).
- Multi-objetive methods: NSGA-II, PAES, ...
- Hybrid algorithms: GASA, CHCES, DEPSO, ...

### *Applications Solved Using MALLBA*

- Natural language tagging.
- Design of combinatorial circuits.
- Bioinformatic problems: DNA fragment assembly and microarray of data.
- Telecommunication problems: radio network design, terminal assignment, sensor networks, VANET, MANET, ...
- Scheduling problems: workforce planning problem, minimum tardy task problem, traffic light controllers, ...
- Continuous benchmark problems: rastrigin function, sphere function, frequency modulation sound problem, Rosenbrock, CEC 2005/2008 standards, GECCO 2009/2010 standards, ...
- Classical optimization problems: travel salesman problem, vehicle route problem, p-median, onemax, maxsat, ...

# B

# Acronyms

| | |
|---|---|
| ACO | Ant Colony Optimization |
| ANOVA | Analysis of Variance |
| cEA | Cellular Evolutionary Algorithm |
| CFLP | Capacitated Facility Location Problem |
| CHC | Cross generational elitist selection, Heterogeneous recombination and Cataclysmic mutation |
| dEA | Distributed Evolutionary Algorithm |
| DOP | Dynamic Optimization Problem |
| DPX | Double Point Crossover |
| EA | Evolutionary Algorithm |
| EP | Evolutionary Programming |
| ES | Evolution Strategy |
| FAP | Fragment Assembly Problem |
| GA | Genetic Algorithm |
| GASA | Hybrid method which combines a genetic algorithm with a simulated annealing |
| GP | Genetic Programming |
| HUX | Half-Uniform Crossover |
| ILS | Iterative Local Search |
| LAN | Local Area Network |
| MAXSAT | Maximum Satisfiability Problem |
| MPI | Message Passing Interface |
| MOGA | Multi-Objective Genetic Algorithm |
| NSGA | Non-Stationary Genetic Algorithm |
| pGA | Parallel Genetic Algorithm |
| PSO | Particle Swarm Optimization |
| TS | Tabu Search |
| SA | Simulated Annealing |
| SS | Scatter Search |
| SPX | Simple Point Crossover |
| SIMD | Single Instruction, Multiple Data |
| UX | Uniform Crossover |
| VNS | Variable Neighborhood Search |
| WAN | Wide Area Network |
| WPP | Workforce Planning Problem |

# References


1. Papadimitriou, C.H., Steiglitz, K.: Combinatorial Optimization: Algorithms and Complexity. Dover Publications, New York (1998)
2. Osman, I.H.: Metastrategy simulated annealing and tabu search algorithms for the vehicle routing problems. Annals of Operations Research 41, 421–451 (1993)
3. Blum, C., Roli, A.: Metaheuristics in combinatorial optimization: Overview and conceptual comparison. ACM Computing Surveys 35(3), 268–308 (2003)
4. Goldberg, D.E.: Genetic Algorithms in Search, Optimization and Machine Learning. Addison-Wesley Publishing Company, Reading (1989)
5. Alba, E., Chicano, J.F., Dorronsoro, B., Luque, G.: Diseño de códigos correctores de errores con algoritmos genéticos. In: Actas del Tercer Congreso Español sobre Metaheurísticas, Algoritmos Evolutivos y Bioinspirados (MAEB), Córdoba, Spain, pp. 51–58 (2004) (Spanish)
6. Alba, E., Tomassini, M.: Parallelism and Evolutionary Algorithms. IEEE Transactions on Evolutionary Computation 6(5), 443–462 (2002)
7. Bäck, T.: Evolutionary Algorithms in Theory and Practice: Evolution Strategies, Evolutionary Programming, Genetic Algorithms. Oxford University Press, New York (1996)
8. Bäck, T., Fogel, D.B., Michalewicz, Z. (eds.): Handbook of Evolutionary Computation. Oxford University Press, Oxford (1997)
9. Cantú-Paz, E.: *Efficient and Accurate Parallel Genetic Algorithms*, 2nd edn. Book Series on Genetic Algorithms and Evolutionary Computation, vol. 1. Kluwer Academic Publishers, Dordrecht (2000)
10. Alba, E. (ed.): Parallel Metaheuristics: A New Class of Algorithms. Wiley, Chichester (2005)
11. Lourenco, H.R., Martin, O., Stützle, T.: Iterated Local Search. In: Handbook of Metaheuristics, pp. 321–353. Kluwer Academic Publishers, Dordrecht (2002)
12. Kirkpatrick, S., Gelatt, C.D., Vecchi, M.P.: Optimization by Simulated Annealing. Science 220, 671–680 (1983)
13. Glover, F., Laguna, M.: Tabu Search. Kluwer Academic Publishers, Dordrecht (1997)
14. Mladenovic, N., Hansen, P.: Variable neighborhood search. Computers & Operations Research 24(11), 1097–1100 (1997)

15. Dorigo, M., Stützle, T.: Ant Colony Optimization. The MIT Press, Cambridge (2004)
16. Darwin, C.: On the Origin of Species by Means of Natural Selection. John Murray, Londres (1859)
17. Mendel, G.: Versuche ber Pflanzen-Hybriden. Verhandlungen des Naturforschedes Vereines in Brnn 4 (1865)
18. Alba, E., Cotta, C.: Evolución de Estructuras de Datos Complejas. Informática y Automática 30(3), 42–60 (1997)
19. Hussain, T.S.: An introduction to evolutionary computation. CITO Researcher Retreat, Hamilton, Ontario (1998)
20. Holland, J.H.: Outline for a logical theory of adaptive systems. Journal of the ACM 9(3), 297–314 (1962)
21. Rechenberg, I.: Cybernetic solution path of an experimental problem. Technical report (1965)
22. Schwefel, H.-P.: Kybernetische Evolution als Strategie der Experimentellen Forschung in der Strmungstechnik. PhD thesis, Technical University of Berlin (1965)
23. Fogel, L.J.: Autonomous automata. Industrial Research 4, 14–19 (1962)
24. Cramer, N.L.: A representation for the adaptive generation of simple sequential programs. In: Grefenstette, J.J. (ed.) Proc. of the First International Conference on Genetic Algorithms and their Applications, Carnegie-Mellon University, Pittsburgh, PA, USA, pp. 183–187. Spartan Books, Washington (1985)
25. Koza, J.R.: Genetic Programming: On the Programming of Computers by Means of Natural Selection. MIT Press, Cambridge (1992)
26. Wright, S.: Isolation by distance. Genetics 28, 114–138 (1943)
27. Alba, E., Troya, J.M.: Improving flexibility and efficiency by adding parallelism to genetic algorithms. Statistics and Computing 12(2), 91–114 (2002)
28. Alba, E.: Análisis y Diseño de Algoritmos Genéticos Paralelos Distribuidos. PhD thesis, Universidad de Málaga (1999)
29. Foster, I., Kesselman, C.: The Grid: Blueprint for a New Computing Infrastructure. Elsevier, Amsterdam (2004)
30. Scott Gordon, V., Darrell Whitley, L.: Serial and parallel genetic algorithms as function optimizers. In: ICGA, pp. 177–183 (1993)
31. Tanese, R.: Distributed genetic algorithms. In: Schaffer, J.D. (ed.) Proceedings of the Third International Conference on Genetic Algorithms (ICGA), pp. 434–439. Morgan Kaufmann, San Francisco (1989)
32. Spiessens, P., Manderick, B.: A massively parallel genetic algorithm: Implementation and first analysis. In: Belew, R.K., Booker, L.B. (eds.) ICGA, pp. 279–287. Morgan Kaufmann, San Francisco (1991)
33. Syswerda, G.: A study of reproduction in generational and steady-state genetic algorithms. In: Rawlins, G.J.E. (ed.) Foundations of Genetic Algorithms, pp. 94–101. Morgan Kaufmann, San Francisco (1991)
34. Levine, D.: Users guide to the PGAPack parallel genetic algorithm library. Technical Report ANL-95/18, Argonne National Laboratory, Mathematics and Computer Science Division, January 31 (1995)
35. Alba, E., Troya, J.M.: A survey of parallel distributed genetic algorithms. Complexity 4(4), 31–52 (1999)
36. Belding, T.C.: The distributed genetic algorithm revisited. In: Eshelman, L.J. (ed.) Proceedings of the Sixth International Conference on Genetic Algorithms (ICGA), pp. 114–121. Morgan Kaufmann, San Francisco (1995)

37. Baluja, S.: Structure and performance of fine-grain parallelism in genetic search. In: Forrest, S. (ed.) Proceedings of the Fifth International Conference on Genetic Algorithms (ICGA), pp. 155–162. Morgan Kaufmann, San Francisco (1993)
38. Maruyama, T., Hirose, T., Konagaya, A.: A fine-grained parallel genetic algorithm for distributed parallel systems. In: Forrest, S. (ed.) Proceedings of the Fifth International Conference on Genetic Algorithms (ICGA), pp. 184–190. Morgan Kaufmann, San Francisco (1993)
39. Tanese, R.: Parallel genetic algorithms for a hypercube. In: Grefenstette, J.J. (ed.) Proceedings of the Second International Conference on Genetic Algorithms (ICGA), p. 177. Lawrence Erlbaum Associates, Mahwah (1987)
40. Lin, S.C., Punch, W.F., Goodman, E.D.: Coarse-grain parallel genetic algorithms: Categorization and a new approach. In: Sixth IEEE Symposium on Parallel and Distributed Processing, pp. 28–37 (1994)
41. Sefrioui, M., Périaux, J.: A hierarchical genetic algorithm using multiple models for optimization. In: Deb, K., Rudolph, G., Lutton, E., Merelo, J.J., Schoenauer, M., Schwefel, H.-P., Yao, X. (eds.) PPSN 2000. LNCS, vol. 1917, pp. 879–888. Springer, Heidelberg (2000)
42. Herrera, F., Lozano, M., Moraga, C.: Hybrid distributed real-coded genetic algorithms. In: Eiben, A.E., Bäck, T., Schoenauer, M., Schwefel, H.-P. (eds.) PPSN 1998. LNCS, vol. 1498, pp. 603–612. Springer, Heidelberg (1998)
43. Herrera, F., Lozano, M.: Gradual distributed real-coded genetic algorithms. IEEE Transactions on Evolutionary Computation 4(1), 43–63 (2000)
44. Luna, F., Alba, E., Nebro, A.J.: Parallel Heterogenous Metaheuristics. In: Alba, E. (ed.) Parallel Metaheuristics, pp. 395–422. Wiley, Chichester (2005)
45. Alba, E., Luna, F., Nebro, A.J., Troya, J.M.: Parallel heterogeneous genetic algorithms for continuous optimization. Parallel Computing 30(5-6), 699–719 (2004), Parallel and nature-inspired computational paradigms and applications
46. Alba, E., Troya, J.M.: Cellular evolutionary algorithms: Evaluating the influence of ratio. In: Deb, K., Rudolph, G., Lutton, E., Merelo, J.J., Schoenauer, M., Schwefel, H.-P., Yao, X. (eds.) PPSN 2000. LNCS, vol. 1917, pp. 29–38. Springer, Heidelberg (2000)
47. Vidal, P., Alba, E.: Cellular Genetic Algorithm on Graphic Processing Units. In: Nature Inspired Cooperative Strategies for Optimization (NICSO 2010), pp. 223–232 (2010)
48. Vidal, P., Alba, E.: A MultiGPU Implementation of a Cellular Genetic Algorithm. In: 2010 IEEE World Congress on Computational Intelligence, IEEE CEC 2010, pp. 1–7 (2010)
49. Gorges-Schleuter, M.: ASPARAGOS an asynchronous parallel genetic optimization strategy. In: Proceedings of the Third International Conference on Genetic Algorithms (ICGA), pp. 422–427. Morgan Kaufmann Publishers Inc., San Francisco (1989)
50. Whitley, D., Starkweather, T.: GENITOR II: A distributed genetic algorithm. Journal of Experimental and Theoretical Aritificial Intelligence 2, 189–214 (1990)
51. Davidor, Y.: A naturally occuring niche and species phenomenon: The model and first results. In: Belew, R.K., Booker, L.B. (eds.) Proceedings of the Fourth International Conference on Genetic Algorithms (ICGA), pp. 257–263 (1991)

52. Mühlenbein, H., Schomish, M., Born, J.: The parallel genetic algorithm as a function optimizer. Parallel Computing 17, 619–632 (1991)
53. Erickson, J.A., Smith, R.E., Goldberg, D.E.: SGA-Cube, a simple genetic algorithm for ncube 2 hypercube parallel computers. Technical Report 91005, The University of Alabama (1991)
54. Robbins, G.: EnGENEer - The evolution of solutions. In: Proceedings of the Fifth Annual Seminar Neural Networks and Genetic Algorithms, London, UK (1992)
55. Stender, J. (ed.): Parallel Genetic Algorithms: Theory and Applications. IOS Press, Amsterdam (1993)
56. Ribeiro-Filho, J.L., Alippi, C., Treleaven, P.: Genetic algorithm programming environments. In: Stender, J. (ed.) Parallel Genetic Algorithms: Theory and Applications, pp. 65–83. IOS Press, Amsterdam (1993)
57. Mejía-Olvera, M., Cantú-Paz, E.: DGENESIS-software for the execution of distributed genetic algorithms. In: Proceedings XX conf. Latinoamericana de Informática, pp. 935–946 (1994)
58. Potts, J.C., Giddens, T.D., Yadav, S.B.: The development and evaluation of an improved genetic algorithm based on migration and artificial selection. IEEE Transactions on Systems, Man, and Cybernetics 24(1), 73–86 (1994)
59. Adamidis, P., Petridis, V.: Co-operating populations with different evolution behavior. In: Proceedings of the Second IEEE Conference on Evolutionary Computation, pp. 188–191. IEEE Press, Los Alamitos (1996)
60. Goodman, E.D.: An Introduction to GALOPPS v3.2. Technical Report 96-07-01, GARAGE, I.S. Lab. Dpt. of C. S. and C.C.C.A.E.M., Michigan State Univ., East Lansing, MI (1996)
61. Talbi, E.-G., Hafidi, Z., Kebbal, D., Geib, J.-M.: MARS: An adaptive parallel programming environment. In: Rajkumar, B. (ed.) High Prformance Cluster Computing, vol. 1, pp. 722–739. Prentice-Hall, Englewood Cliffs (1999)
62. Radcliffe, N.J., Surry, P.D.: The reproductive plan language RPL2: Motivation, architecture and applications. In: Stender, J., Hillebrand, E., Kingdon, J. (eds.) Genetic Algorithms in Optimisation, Simulation and Modelling. IOS Press, Amsterdam (1999)
63. Arenas, M., Collet, P., Eiben, A.E., Jelasity, M., Merelo, J.J., Paechter, B., Preuß, M., Schoenauer, M.: A framework for distributed evolutionary algorithms. In: Guervós, J.J.M., Adamidis, P.A., Beyer, H.-G., Fernández-Villacañas, J.-L., Schwefel, H.-P. (eds.) PPSN 2002. LNCS, vol. 2439, pp. 665–675. Springer, Heidelberg (2002)
64. Alba, E., Almeida, F., Blesa, M., Cotta, C., Díaz, M., Dorta, I., Gabarró, J., León, C., Luque, G., Petit, J.: Rodríguez C., Rojas A., and Xhafa F. Efficient parallel LAN/WAN algorithms for optimization. The MALLBA project. Parallel Computing 32(5-6), 415–440 (2006)
65. Cahon, S., Melab, N., Talbi, E.-G.: ParadisEO: A Framework for the Reusable Design of Parallel and Distributed Metaheuristics. Journal of Heuristics 10(3), 357–380 (2004)
66. Helseth, A., Holen, A.T.: Impact of Energy End Use and Customer Interruption Cost on Optimal Allocation of Switchgear in Constrained Distribution Networks. IEEE Transactions on Power Delivery 23(3), 1419–1425 (2008)

67. Luque, G., Alba, E., Dorronsoro, B.: An Asynchronous Parallel Implementation of a Cellular Genetic Algorithm for Combinatorial Optimization. In: Genetic and Evolutionary Computation Conference (GECCO 2009), pp. 1395–1401. IEEE Press, Montreal (2009)
68. Alba, E., Troya, J.M.: Gaining new fields of application for OOP: the parallel evolutionary algorithm case. Journal of Object Oriented Programming (December 2001) (web version only)
69. Alba, E., Saucedo, J.F., Luque, G.: A Study of Canonical GAs for NSOPs. In: Alba, E., Saucedo, J.F., Luque, G. (eds.) MIC 2005 Post Conference Volume on Metaheuristics - Progress in Complex Systems Optimization, ch. 13, pp. 245–260. Springer, Heidelberg (2007)
70. Alba, E., Luque, G., Arias, D.: Impact of Frequency and Severity on Non-Stationary Optimization Problems. In: 6th European Workshop on Evolutionary Algorithms in Stochastic and Dynamic Environments (EvoSTOC 2009), Tubingen, Germany, pp. 755–761. Springer, Heidelberg (2009)
71. Luque, G., Alba, E., Dorronsoro, B.: Selection Pressure and Takeover Time of Distributed Evolutionary Algorithms. In: Genetic and Evolutionary Computation Conference (GECCO 2010), pp. 1083–1088. IEEE Press, Portland (2010)
72. Chicano, F., Luque, G., Alba, E.: Elementary Landscape Decomposition of the Quadratic Assignment Problem. In: Genetic and Evolutionary Computation Conference (GECCO 2010), pp. 1425–1432. IEEE Press, Portland (2010)
73. Whitley, D., Chicano, F., Alba, E., Luna, F.: Elementary Landscapes of Frequency Assignment Problems. In: Genetic and Evolutionary Computation Conference (GECCO 2010), pp. 1409–1416. IEEE Press, Portland (2010)
74. Oram, A.: Peer-to-peer: Harnessing the power of disruptive technologies (2001)
75. Van-Luong, T., Melab, N., Talbi, E.-G.: GPU-based Island Model for Evolutionary Algorithms. In: Genetic and Evolutionary Computation Conference (GECCO 2010), pp. 1089–1096. IEEE Press, Portland (2010)
76. Wilson, G., Banzhaf, W.: Deployment of cpu and gpu-based genetic programming on heterogeneous devices. In: GECCO 2009: Proceedings of the 11th Annual Conference Companion on Genetic and Evolutionary Computation Conference, pp. 2531–2538. ACM, New York (2009)
77. Tsutsui, S., Fujimoto, N.: Solving quadratic assignment problems by genetic algorithms with gpu computation: a case study. In: GECCO 2009: Proceedings of the 11th Annual Conference Companion on Genetic and Evolutionary Computation Conference, pp. 2523–2530. ACM, New York (2009)
78. Garey, M.R., Johnson, D.S.: Computers and Intractability. A guide to the Theory of NP-Completeness. Freeman, San Francisco (1979)
79. De Jong, K.A., Potter, M.A., Spears, W.M.: Using problem generators to explore the effects of epistasis. In: 7th ICGA, pp. 338–345. Kaufman, San Francisco (1997)
80. Graham, R.L.: Bounds on multiprocessor timing anomalies. SIAM Journal of Applied Mathematics 17, 416–429 (1969)
81. Karp, R.M.: Probabilistic analysis of partitioning algorithms for the traveling salesman problem in the plane. Mathematics of Operations Research 2, 209–224 (1977)
82. Barr, R.S., Hickman, B.L.: Reporting Computational Experiments with Parallel Algorithms: Issues, Measures, and Experts' Opinions. ORSA Journal on Computing 5(1), 2–18 (1993)

83. Rardin, R.L., Uzsoy, R.: Experimental Evaluation of Heuristic Optimization Algorihtms: A Tutorial. Journal of Heuristics 7(3), 261–304 (2001)
84. McGeogh, C.: Toward an experimental method for algorithm simulation. INFORMS Journal on Computing 8(1) (1995)
85. Alba, E.: Parallel evolutionary algorithms can achieve super-linear performace. Information Processing Letters 82, 7–13 (2002)
86. UEA CALMA Group. Calma project report 2.4: Parallelism in combinatorial optimisation. Technical report, School of Information Systems, University of East Anglia, Norwich, UK, September 18 (1995)
87. Alba, E., Nebro, A.J., Troya, J.M.: Heterogeneous Computing and Parallel Genetic Algorithms. Journal of Parallel and Distributed Computing 62, 1362–1385 (2002)
88. Donalson, V., Berman, F., Paturi, R.: Program speedup in heterogeneous computing network. Journal of Parallel and Distributed Computing 21, 316–322 (1994)
89. Karp, A.H., Flatt, H.P.: Measuring Parallel Processor Performance. Communications of the ACM 33, 539–543 (1990)
90. Hooker, J.N.: Testing heuristics: We have it all wrong. Journal of Heuristics 1(1), 33–42 (1995)
91. Eiben, A.E., Jelasity, M.: A critical note on experimental reseeach methodology in ec. In: Congress on Evolutionary Computation 2002, pp. 582–587. IEEE Press, Los Alamitos (2002)
92. Darrell Whitley, L.: An overview of evolutionary algorithms: practical issues and common pitfalls. Information & Software Technology 43(14), 817–831 (2001)
93. Reinelt, G.: TSPLIB - A travelling salesman problem library. ORSA - Jorunal of Computing 3, 376–384 (1991)
94. Uzsoy, R., Demirkol, E., Mehta, S.V.: Benchmarks for shop scheduling problems. European Journal of Operational Research 109, 137–141 (1998)
95. Finck, S., Hansen, N., Ros, R., Auger, A.: Real-parameter black-box optimization benchmarking 2009: Presentation of the noiseless functions. Technical Report Technical Report 2009/20, Research Center PPE (2009)
96. Suganthan, P.N., Hansen, N., Liang, J.J., Deb, K., Chen, Y.-P., Auger, A., Tiwari, S.: Problem definitions and evaluation criteria for the cec 2005 special session on real-parameter optimization. Technical Report KanGAL Report 2005005, Nanyang Technological University, IIT Kanpur, India (May 2005)
97. Tang, K., Yao, X., Suganthan, P.N., MacNish, C., Chen, Y.P., Chen, C.M., Yang, Z.: Benchmark functions for the cec 2008 special session and competition on large scale global optimization. Technical report, Nature Inspired Computation and Applications Laboratory, USTC, China (2007)
98. Beasley, J.E.: OR-library: distributing test problems by electronic mail. Journal of the Operational Research Society 41(11), 1069–1072 (1990)
99. Montgomery, D.C.: Design and Analysis of Experiments, 3rd edn. John Wiley, New York (1991)
100. Birattari, M., Stützle, T., Paquete, L., Varrentrapp, K.: A racing algorithm for configuring metaheuristics. In: GECCO, vol. 2, pp. 11–18 (2002)
101. Bartz-Beielstein, T., Preuss, M.: The future of experimental research. In: GECCO 2009: Proceedings of the 11th Annual Conference Companion on Genetic and Evolutionary Computation Conference, pp. 3185–3226. ACM, New York (2009)

102. Golden, B., Stewart, W.: Empirical Analisys of Heuristics. In: Lawlwer, E., Lenstra, J., Rinnooy Kan, A., Schoys, D. (eds.) The Traveling Salesman Problem, a Guided Tour of Combinatorial Optimization, pp. 207–249. Wiley, Chichester (1985)
103. Cleveland, W.S.: Elements of Graphing Data. Wadsworth, Monteray (1985)
104. Tufte, E.R.: The Visual Display of Quantitative Information. Graphics Press (1993)
105. Alcalá-Fdez, J., Sánchez, L., García, S., del Jesus, M.J., Ventura, S., Garrell, J.M., Otero, J., Romero, C., Bacardit, J., Rivas, V.M., et al.: KEEL: a software tool to assess evolutionary algorithms for data mining problems. Soft Computing-A Fusion of Foundations, Methodologies and Applications 13(3), 307–318 (2009)
106. Andre, D., Koza, J.R.: Parallel genetic programming: A scalable implementation using the transputer network architecture. In: Angeline, P.J., Kinnear Jr., K.E. (eds.) Advances in Genetic Programming 2, ch. 16, pp. 317–338. MIT Press, Cambridge (1996)
107. Alba, E., Troya, J.M.: Influence of the Migration Policy in Parallel dGAs with Structured and Panmictic Populations. Applied Intelligence 12(3), 163–181 (2000)
108. Sarma, J., De Jong, K.A.: An analysis of the effect of the neighborhood size and shape on local selection algorithms. In: Ebeling, W., Rechenberg, I., Voigt, H.-M., Schwefel, H.-P. (eds.) PPSN 1996. LNCS, vol. 1141, pp. 236–244. Springer, Heidelberg (1996)
109. Sarma, J., De Jong, K.: An Analysis of Local Selection Algorithms in a Spatially Structured Evolutionary Algorithm. In: Bäck, T. (ed.) Proceedings of the 7th International Conference on Genetic Algorithms, pp. 181–186. Morgan Kaufmann, San Francisco (1997)
110. Gorges-Schleuter, M.: An Analysis of Local Selection in Evolution Strategies. In: Banzhaf, W., Daida, J., Eiben, A.E., Garzon, M.H., Honavar, V., Jakiela, M., Smith, R.E. (eds.) Proceedings of the Genetic and Evolutionary Computation Conference, vol. 1, pp. 847–854. Morgan Kaufmann, Orlando (1999)
111. Rudolph, G.: Takeover Times in Spatially Structured Populations: Array and Ring. In: Lai, K.K., Katai, O., Gen, M., Lin, B. (eds.) 2nd Asia-Pacific Conference on Genetic Algorithms and Applications, pp. 144–151. Global-Link Publishing (2000)
112. Giacobini, M., Tettamanzi, A., Tomassini, M.: Modelling Selection Intensity for Linear Cellular Evolutionary Algorithms. In: Liardet, P., et al. (eds.) Artificial Evolution, Sixth International Conference, pp. 345–356. Springer, Heidelberg (2003)
113. Giacobini, M., Alba, E., Tomassini, M.: Selection Intensity in Asynchronous Cellular Evolutionary Algorithms. In: Cantú-Paz, E. (ed.) Proceedings of the Genetic and Evolutionary Computation Conference, Chicago, USA, pp. 955–966 (2003)
114. Sprave, J.: A Unified Model of Non-Panmictic Population Structures in Evolutionary Algorithms. In: Angeline, P.J., Michalewicz, Z., Schoenauer, M., Yao, X., Zalzala, A. (eds.) Proceedings of the Congress of Evolutionary Computation, vol. 2, pp. 1384–1391. IEEE Press, Mayflower Hotel (1999)
115. Alba, E., Luque, G.: Growth Curves and Takeover Time in Evolutionary Algorithms. In: Deb, K., et al. (eds.) Genetic and Evolutionary Computation Conference (GECCO 2004), Seattle, Washington, pp. 864–876 (2004)

116. Goldberg, D.E., Deb, K.: A comparative analysis of selection schemes used in genetic algorithms. In: Rawlins, G.J.E. (ed.) Foundations of Genetic Algorithms, pp. 69–93. Morgan Kaufmann, San Francisco (1991)
117. Chakraborty, U.K., Deb, K., Chakraborty, M.: Analysis of Selection Algorithms: A Markov Chain Approach. Evolutionary Computation 4(2), 133–167 (1997)
118. Giacobini, M., Tomassini, M., Tettamanzi, A., Alba, E.: The Selection Intensity in Cellular Evolutionary Algorithms for Regular Lattices. IEEE Transactions on Evolutionary Computation 5(9), 489–505 (2005)
119. Manning, C.D., Schütze, H.: Foundations of Statistical Natural Language Processing. MIT Press, Cambridge (2000)
120. Pla, F., Molina, A., Prieto, N.: Tagging and chunking with bigrams. In: Proc. of the 17th Conference on Computational Linguistics, pp. 614–620 (2000)
121. Baeza-Yates, R.A., Ribeiro-Neto, B.A.: Modern Information Retrieval. Addison-Wesley, Reading (1999)
122. DeRose, S.J.: Grammatical Category Disambiguation by Statistical Optimization. Computational Linguistics 14, 31–39 (1988)
123. Nelson, F.W., Kucera, H.: Manual of information to accompany a standard corpus of present-day edited american english, for use with digital computers. Technical report, Dep. of Linguistics, Brown University (1979)
124. Charniak, E.: Statistical Language Learning. MIT Press, Cambridge (1993)
125. Forney, G.D.: The viterbi algorithm. Proceedings of The IEEE 61(3), 268–278 (1973)
126. Araujo, L.: Part-of-speech tagging with evolutionary algorithms. In: Gelbukh, A. (ed.) CICLing 2002. LNCS, vol. 2276, pp. 230–239. Springer, Heidelberg (2002)
127. Araujo, L., Luque, G., Alba, E.: Metaheuristics for Natural Language Tagging. In: Deb, K., et al. (eds.) GECCO 2004. LNCS, vol. 3102, pp. 889–900. Springer, Heidelberg (2004)
128. Alba, E., Luque, G., Araujo, L.: Natural language tagging with genetic algorithms. Information Processing Letters 100(5), 173–182 (2006)
129. Erik, F.: Tjong Kim Sang. Memory-based shallow parsing. J. Mach. Learn. Res. 2, 559–594 (2002)
130. Davis, L. (ed.): Handbook of Genetic Algorithms. Van Nostrand Reinhold, New York (1991)
131. Eshelman, L.J.: The CHC Adaptive Search Algorithm: How to Have Safe Search when Engaging in Nontraditional Genetic Recombination. In: Rawlins, G.E. (ed.) FOGA, pp. 265–283. Morgan Kaufmann, San Francisco (1991)
132. Sampson, G.: English for the Computer. Clarendon Press, Oxford (1995)
133. Brants, T.: Tnt: a statistical part-of-speech tagger. In: Proceedings of the Sixth Conference on Applied Natural Language Processing, pp. 224–231. Morgan Kaufmann Publishers Inc., San Francisco (2000)
134. Marcus, M.P., Santorini, B., Marcinkiewicz, M.A.: Building a large annotated corpus of english: The penn treebank. Computational Linguistics 19(2), 313–330 (1994)
135. Pla, F., Molina, A.: Improving part-of-speech tagging using lexicalized hmms. Nat. Lang. Eng. 10(2), 167–189 (2004)

136. van Halteren, H., Daelemans, W., Zavrel, J.: Improving accuracy in word class tagging through the combination of machine learning systems. Comput. Linguist. 27(2), 199–229 (2001)
137. Karnaugh, M.: A map method for synthesis of combinational logic circuits. Transactions of the AIEE, Communications and Electronics I(72), 593–599 (1953)
138. Veitch, E.W.: A chart method for simplifying boolean functions. In: Proceedings of the ACM, pp. 127–133. IEEE Service Center, Piscataway (1952)
139. McCluskey, E.J.: Minimization of boolean functions. Bell Systems Technical Journal 35(5), 1417–1444 (1956)
140. Quine, W.V.: A way to simplify truth functions. American Mathematical Monthly 62(9), 627–631 (1955)
141. Coello Coello, C.A., Christiansen, A.D., Hernández Aguirre, A.: Use of Evolutionary Techniques to Automate the Design of Combinational Circuits. International Journal of Smart Engineering System Design 2(4), 299–314 (2000)
142. Miller, J.F., Job, D., Vassilev, V.K.: Principles in the Evolutionary Design of Digital Circuits—Part I. Genetic Programming and Evolvable Machines 1(1/2), 7–35 (2000)
143. Pérez, E.I., Coello Coello, C.A., Aguirre, A.H.: Extracting and re-using design patterns from genetic algorithms using case-based reasoning. Engineering Optimization 35(2), 121–141 (2003)
144. Miller, J., Kalganova, T., Lipnitskaya, N., Job, D.: The Genetic Algorithm as a Discovery Engine: Strange Circuits and New Principles. In: Proceedings of the AISB Symposium on Creative Evolutionary Systems (CES 1999), Edinburgh, UK (1999)
145. Alba, E., Luque, G., Coello, C., Hernández, E.: Comparative study of serial and parallel heuristics used to design combinational logic circuits. Optimization Methods and Software 22(3), 485–509 (2007)
146. Coello Coello, C.A., Hernández Aguirre, A., Buckles, B.P.: Evolutionary Multiobjective Design of Combinational Logic Circuits. In: Lohn, J., Stoica, A., Keymeulen, D., Colombano, S. (eds.) Proceedings of the Second NASA/DoD Workshop on Evolvable Hardware, pp. 161–170. IEEE Computer Society, Los Alamitos (2000)
147. Coello Coello, C.A., Christiansen, A.D., Hernández Aguirre, A.: Automated Design of Combinational Logic Circuits using Genetic Algorithms. In: Smith, D.G., Steele, N.C., Albrecht, R.F. (eds.) Proceedings of the International Conference on Artificial Neural Nets and Genetic Algorithms, pp. 335–338. Springer, University of East Anglia (1997)
148. Louis, S.J.: Genetic Algorithms as a Computational Tool for Design. PhD thesis, Department of Computer Science, Indiana University (August 1993)
149. Louis, S.J., Rawlins, G.J.: Using Genetic Algorithms to Design Structures. Technical Report 326, Computer Science Department, Indiana University, Bloomington, Indiana (February 1991)
150. Miller, J.F., Thomson, P., Fogarty, T.: Designing Electronic Circuits Using EAs. Arithmetic Circuits: A Case Study. In: Quagliarella, D., Périaux, J., Poloni, C., Winter, G. (eds.) Genetic Algorithms and Evolution Strategy in Engineering and Computer Science, pp. 105–131. Morgan Kaufmann, San Francisco (1998)

151. Banzhaf, W., Nordin, P., Keller, R.E., Francone, F.D.: Genetic Programming. An Introduction. On the Automatic Evolution of Computer Programs and Its Applications. Morgan Kaufmann Publishers, San Francisco (1998)
152. Al-Saiari, U.S.: Digital circuit design through simulated evolution. Master's thesis, King Fahd University of Petroleum and Minerals, Dhahran, Saudi Arabia (November 2003)
153. Słowik, A., Białko, M.: Design and Optimization of Combinational Digital Circuits Using Modified Evolutionary Algorithm. In: Rutkowski, L., Siekmann, J.H., Tadeusiewicz, R., Zadeh, L.A. (eds.) ICAISC 2004. LNCS (LNAI), vol. 3070, pp. 468–473. Springer, Heidelberg (2004)
154. Vassilev, V.K., Miller, J.F., Fogarty, T.C.: Digital Circuit Evolution and Fitness Landscapes. In: 1999 Congress on Evolutionary Computation, vol. 2, pp. 1299–1306. IEEE Service Center, Piscataway (1999)
155. Miller, J.F., Job, D., Vassilev, V.K.: Principles in the Evolutionary Design of Digital Circuits—Part II. Genetic Programming and Evolvable Machines 1(3), 259–288 (2000)
156. Vassilev, V.K., Fogarty, T.C., Miller, J.F.: Information Characteristics and the Structure of Landscapes. Evolutionary Computation 8(1), 31–60 (2000)
157. Abd-El-Barr, M., Sait, S.M., Sarif, B.A.B., Al-Saiari, U.: A modified ant colony algorithm for evolutionary design of digital circuits. In: Proceedings of the 2003 Congress on Evolutionary Computation (CEC 2003), pp. 708–715. IEEE Press, Canberra (2003)
158. Coello Coello, C.A., Zavala Gutiérrez, R.L., García, B.M., Aguirre, A.H.: Ant Colony System for the Design of Combinational Logic Circuits. In: Miller, J., Thompson, A., Thomson, P., Fogarty, T.C. (eds.) Evolvable Systems: From Biology to Hardware, pp. 21–30. Springer, Edinburgh (2000)
159. Coello Coello, C.A., Hernández Luna, E., Hernández Aguirre, A.: Use of Particle Swarm Optimization to Design Combinational Logic Circuits. In: Tyrrell, A.M., Haddow, P.C., Torresen, J. (eds.) ICES 2003. LNCS, vol. 2606, pp. 398–409. Springer, Heidelberg (2003)
160. Venu, G.: Gudise and Ganesh K. Venayagamoorthy. Evolving digital circuits using particle swarm. In: Proceedings of the INNS-IEEE International Joint Conference on Neural Networks, Porland, OR, USA, pp. 468–472 (2003)
161. Gordon, T.G.W., Bentley, P.J.: On evolvable hardware. In: Ovaska, S., Sztandera, L. (eds.) Soft Computing in Industrial Electronics, pp. 279–323. Physica-Verlag, Heidelberg (2003)
162. Holland, J.H.: Adaptation in Natural and Artificial Systems. The University of Michigan Press, Ann Arbor (1975)
163. Aarts, E., Korst, J.: Selected topics in simulated annealing. In: Ribero, C.C., Hansen, P. (eds.) Essays and Surveys un Metaheuristics. Kluwer Academic Publishers, Boston (2002)
164. Laarhoven, P.J.M., Aarts, E.H.L. (eds.): Simulated annealing: theory and applications. Kluwer Academic Publishers, Norwell (1987)
165. Chelouah, R., Siarry, P.: Tabu search applied to global optimization. European Journal of Operational Research 123(2), 256–270 (2000)
166. Fan, S.-K., Liang, Y.-C., Zahara, E.: Hybrid simplex search and particle swarm optimization for the global optimization of multimodal functions. Engineering Optimization 36(4), 401–418 (2004)

167. Hedar, A.-R., Fukushima, M.: Hybrid simulated annealing and direct search method for nonlinear unconstrained global optimization. Optimization Methods and Software 17(5), 891–912 (2002)
168. Hedar, A.-R., Fukushima, M.: Heuristic pattern search and its hybridization with simulated annealing for nonlinear global optimization. Optimization Methods and Software 19(3-4), 291–308 (2004)
169. Kvasnièka, V., Pospíchal, J.: A hybrid of simplex method and simulated annealing. Chemometrics and Intelligent Laboratory Systems 39, 161–173 (1997)
170. Serna Pérez, E.: Diseño de Circuitos Lógicos Combinatorios utilizando Programación Genética. Master's thesis, Maestría en Inteligencia Artificial, Facultad de Física e Inteligencia Artificial, Universidad Veracruzana, Enero 2001 (2001) (in Spanish)
171. Luna, E.H.: Diseño de circuitos lógicos combinatorios usando optimización mediante cúmulos de partículas. Master's thesis, Computer Science Section, Electrical Engineering Department, CINVESTAV-IPN, Mexico, D.F., Mexico (February 2004) (in Spanish)
172. Sasao, T. (ed.): Logic Synthesis and Optimization. Kluwer Academic Press, Dordrecht (1993)
173. Coello Coello, C.A., Aguirre, A.H.: Design of combinational logic circuits through an evolutionary multiobjective optimization approach. Artificial Intelligence for Engineering, Design, Analysis and Manufacture 16(1), 39–53 (2002)
174. Alba, E., Laguna, M., Luque, G.: Workforce planning with a parallel genetic algorithm. In: Arenas, M.G., Herrera, F., Lozano, M., Merelo, J.J., Romero, G., Sánchez, A.M. (eds.) IV Congreso Español de Metaheurísticas, Algoritmos Evolutivos y Bioinspirados (MAEB 2005 - CEDI 2005), Granada, España, pp. 919–919 (2005)
175. Alba, E., Luque, G., Luna, F.: Parallel metaheuristics for workforce planning. In: NIDISC 2006-IPDPS 2006, p. 246. IEEE Press, Los Alamitos (2006)
176. Glover, F., Kochenberger, G., Laguna, M., Wubbena, T.: Selection and Assignment of a Skilled Workforce to Meet Job Requirements in a Fixed Planning Period. In: MAEB 2004, pp. 636–641 (2004)
177. Laguna, M., Wubbena, T.: Modeling and Solving a Selection and Assignment Problem. In: Golden, B.L., Raghavan, S., Wasil, E.A. (eds.) The Next Wave in Computing, Optimization, and Decision Technologies, pp. 149–162. Springer, Heidelberg (2005)
178. Aardal, K.: Capacitated Facility Location: Separation Algorithm and Computational Experience. Mathematical Programming 81, 149–175 (1998)
179. Klose, A.: An LP-based Heuristic for Two-stage Capacitated Facility location Problems. Journal of the Operational Research Society 50, 157–166 (1999)
180. Glover, F., Laguna, M., Martí, R.: Fundamentals of scatter search and path relinking. Control and Cybernetics 39(3), 653–684 (2000)
181. García-López, F., Melián-Batista, B., Moreno-Pérez, J., Moreno-Vega, J.M.: Parallelization of the Scatter Search. Parallel Computing 29, 575–589 (2003)
182. Nebro, A.J., Luque, G., Luna, F., Alba, E.: DNA fragment assembly using a grid-based genetic algorithm. Computers & Operations Research 35(9), 2776–2790 (2008)
183. Minetti, G., Alba, E., Luque, G.: Seeding Strategies and Recombination Operators for Solving the DNA Fragment Assembly Problem. Information Processing Letters 108(3), 94–100 (2008)

184. Parsons, R., Forrest, S., Burks, C.: Genetic algorithms, operators, and DNA fragment assembly. Machine Learning 21, 11–33 (1995)
185. Pevzner, P.A.: Computational molecular biology: An algorithmic approach. The MIT Press, London (2000)
186. Setubal, J., Meidanis, J.: Fragment Assembly of DNA. In: Introduction to Computational Molecular Biology, ch. 4, pp. 105–139. University of Campinas, Brazil (1997)
187. Kim, S.: A structured Pattern Matching Approach to Shotgun Sequence Assembly. PhD thesis, Computer Science Department, The University of Iowa, Iowa City (1997)
188. Allex, C.F.: Computational Methods for Fast and Accurate DNA Fragment Assembly. UW technical report CS-TR-99-1406, Department of Computer Sciences, University of Wisconsin-Madison (1999)
189. Notredame, C., Higgins, D.G.: SAGA: sequence alignment by genetic algorithm. Nucleic Acids Research 24, 1515–1524 (1996)
190. Parsons, R., Johnson, M.E.: A case study in experimental design applied to genetic algorithms with applications to DNA sequence assembly. American Journal of Mathematical and Management Sciences 17, 369–396 (1995)
191. Green, P.: Phrap, `http://www.mbt.washington.edu/phrap.docs/phrap.html`
192. Sutton, G.G., White, O., Adams, M.D., Kerlavage, A.R.: TIGR Assembler: A new tool for assembling large shotgun sequencing projects. In: Genome Science & Technology, pp. 9–19 (1995)
193. Chen, T., Skiena, S.S.: Trie-based data structures for sequence assembly. In: The Eighth Symposium on Combinatorial Pattern Matching, pp. 206–223 (1998)
194. Huang, X., Madan, A.: CAP3: A DNA sequence assembly program. Genome Research 9, 868–877 (1999)
195. Myers, E.W.: Towards simplifying and accurately formulating fragment assembly. Journal of Computational Biology 2(2), 275–290 (2000)
196. Alba, E., Luque, G.: A new local search algorithm for the dna fragment assembly problem. In: Cotta, C., van Hemert, J. (eds.) EvoCOP 2007. LNCS, vol. 4446, pp. 1–12. Springer, Heidelberg (2007)
197. Luque, G., Alba, E.: Metaheuristics for the DNA Fragment Assembly Problem. International Journal of Computational Intelligence Research 1(1-2), 98–108 (2005)
198. Minetti, G., Alba, E., Luque, G.: Variable neighborhood search for solving the dna fragment assembly problem. In: CACIC 2007, Corrientes y Resistencia, Argentina (October 2007)
199. Minetti, G., Alba, E., Luque, G.: Variable Neighborhood Search as Genetic Operator for DNA Fragment Assembling Problem. In: Eighth International Conference on Hybrid Intelligent System (HIS 2008), pp. 714–719. IEEE Press, Barcelona (2008)
200. Whitely, D.: The GENITOR algorithm and selection pressure: Why rank-based allocation of reproductive trials is best. In: Schaffer, J.D. (ed.) Proceedings of the Third International Conference on Genetic Algorithms, pp. 116–121. Morgan Kaufmann, San Francisco (1989)
201. Engle, M.L., Burks, C.: Artificially generated data sets for testing DNA fragment assembly algorithms. Genomics 16 (1993)

202. Li, L., Khuri, S.: A comparison of dna fragment assembly algorithms. In: International Conference on Mathematics and Engineering Techniques in Medicine and Biological Sciences, pp. 329–335 (2004)
203. Minetti, G., Alba, E.: Metaheuristic Assemblers of DNA strands: Noiseless and Noisy Cases. In: 2010 IEEE World Congress on Computational Intelligence, IEEE CEC 2010, pp. 1–8 (2010)
204. Alba, E., Luque, G., Garcia-Nieto, J., Ordóñez, G., Leguizamón, G.: MALLBA: a software library to design efficient optimisation algorithms. International Journal of Innovative Computing and Applications 1(1), 74–85 (2007)
205. Gamma, E., Helm, R., Johnson, R., Vlissides, J.: Design Patterns: Elements of Reusable Object-Oriented Software. Addison-Wesley, Reading (1995)
206. Alba, E., Cotta, C., Díaz, M., Soler, E., Troya, J.M.: MALLBA: Middleware for a geographically distributed optimization system. Technical report, Dpto. Lenguajes y Ciencias de la Computación, Universidad de Málaga, internal report (2000)
207. Cotta, C., Troya, J.M.: On Decision-Making in Strong Hybrid Evolutionary Algorithms. In: Mira, J., Moonis, A., de Pobil, A.P. (eds.) IEA/AIE 1998. LNCS, vol. 1416, pp. 418–427. Springer, Heidelberg (1998)
208. Wolpert, D.H., Macready, W.G.: No free lunch theorems for optimization. IEEE Transactions on Evolutionary Computation 1(1), 67–82 (1997)
209. Daida, J.M., Ross, S.J., Hannan, B.C.: Biological Symbiosis as a Metaphor for Computational Hybridization. In: Eshelman, L.J. (ed.) Sixth International Conference on Genetic Algorithms, pp. 328–335. Morgan Kaufmann, San Francisco (1995)

CPSIA information can be obtained at www.ICGtesting.com
Printed in the USA

241012LV00003B/18/P

9 783642 220838